Climate Change and Genocide

Climate change caused by human activity is the most fundamental challenge facing mankind in the 21st century, since it will drastically alter the living conditions of millions of people, mainly in the Global South. Environmental violence, including resource crises such as peak fossil fuel, will lie at the heart of future conflicts. However, Genocide Studies have so far neglected this subject, due to the emphasis that traditional genocide scholarship places on ideology and legal prosecution, leading to a narrow understanding of the driving forces of genocide.

This books aims at changing this, initiating a dialogue between scholars working in the areas of climate change and genocide. Research into genocide as well as climate change is a highly interdisciplinary endeavour, transcending the boundaries of established disciplines. Contributions to this book address this by approaching the subject from a wide array of methodological, theoretical, disciplinary and regional perspectives. As all the contributions show, climate change is a major threat multiplier for violence or non-violent destruction and any understanding of prevention needs to take this into account. They offer a basis for much needed Critical Prevention Studies, which aims at sustainable prevention.

This book was originally published as a special issue of *The International Journal of Human Rights.*

Jürgen Zimmerer is Professor of History at the University of Hamburg, Germany, and President of the International Network of Genocide Scholars. His research interests include German Colonialism, Comparative Genocide, Colonialism and the Holocaust, Critical Prevention Studies and Environmental Violence and Genocide. He is the author and editor of nine books and numerous articles.

Climate Change and Genocide

Environmental Violence in the 21st Century

Edited by

Jürgen Zimmerer

Routledge
Taylor & Francis Group
LONDON AND NEW YORK

First published 2015 by Routledge

2 Park Square, Milton Park, Abingdon, Oxon OX14 4RN
711 Third Avenue, New York, NY 10017, USA

Routledge is an imprint of the Taylor & Francis Group, an informa business

First issued in paperback 2017

British Library Cataloguing in Publication Data
A catalogue record for this book is available from the British Library

ISBN 13: 978-1-138-88642-1 (hbk)
ISBN 13: 978-1-138-05889-7 (pbk)

Typeset in Times New Roman
by RefineCatch Limited, Bungay, Suffolk

Publisher's Note
The publisher accepts responsibility for any inconsistencies that may have arisen during the conversion of this book from journal articles to book chapters, namely the possible inclusion of journal terminology.

Contents

Citation Information

The chapters in this book were originally published in *The International Journal of Human Rights*, volume 18, issue 3 (March 2014). When citing this material, please use the original page numbering for each article, as follows:

Foreword

Foreword
Jürgen Zimmerer
The International Journal of Human Rights, volume 18, issue 3 (March 2014)
pp. 263–264

Chapter 1

Climate change, environmental violence and genocide
Jürgen Zimmerer
The International Journal of Human Rights, volume 18, issue 3 (March 2014)
pp. 265–280

Chapter 2

Subsistence societies, globalisation, climate change and genocide: discourses of vulnerability and resilience
Mark Levene and Daniele Conversi
The International Journal of Human Rights, volume 18, issue 3 (March 2014)
pp. 281–297

Chapter 3

Marx, Lemkin and the genocide–ecocide nexus
Martin Crook and Damien Short
The International Journal of Human Rights, volume 18, issue 3 (March 2014)
pp. 298–319

Chapter 4

Crystallisations of the global western state in the era of climate change
Gregory Kent
The International Journal of Human Rights, volume 18, issue 3 (March 2014)
pp. 320–335

Chapter 5

Culturecide in changing Micronesian climates? About the unintenionality of climate change
Rebecca Hofmann
The International Journal of Human Rights, volume 18, issue 3 (March 2014) pp. 336–349

Chapter 6

Genocidal risk and climate change: Africa in the twenty-first century
Andreas Exenberger and Andreas Pondorfer
The International Journal of Human Rights, volume 18, issue 3 (March 2014) pp. 350–368

Chapter 7

Violent climate or climate of violence? Concepts and relations with focus on Kenya and Sudan
Jürgen Scheffran, Tobias Ide and Janpeter Schilling
The International Journal of Human Rights, volume 18, issue 3 (March 2014) pp. 369–390

Please direct any queries you may have about the citations to clsuk.permissions@cengage.com

Notes on Contributors

Daniele Conversi is a social historian and political theorist best known for his comparative work on nationalism, ethno-cultural diversity and the politics of culture, particularly in Spain and the Mediterranean. He has written extensively on the interaction between ideologies of modernity, conflict, war, genocide and globalization, more recently focusing on the cultural foundations of the nation-state. He is currently Research Professor at the University of the Basque Country, Bilbao, Spain and the Ikerbasque Foundation for Science.

Martin Crook is a Research Associate and PhD candidate at the Human Rights Consortium (HRC), School of Advanced Study, University of London, UK.

Andreas Exenberger is Assistant Professor of Economic and Social History at the Leopold-Franzens- University of Innsbruck, Austria, and speaker of the cluster 'World order and extremes' of the research platform 'Religion–politics–art: platform for conflict and communication research'. His main interests are globalization history, hunger and violence, with an increasing focus on Africa, especially the Congo region and the Horn of Africa. His publications include the edited volumes *Afrika: Kontinent der Extreme?* (2011) and *Facetten der Umweltkrise* (2010). He is vice president of the Society for Communication and Development in Salzburg.

Rebecca Hofmann studied cultural anthropology, geography and Spanish philology at the University of Freiburg, Germany, and in Seville, Spain. After working in Hyderabad, India, as a member of the German Federal Ministry of Education and Research-funded project entitled 'Research for the Sustainable Development of the Megacities of Tomorrow: Energy and Climate-efficient Structures in Urban Growth Centers,' she is now part of the Climates of Migration project at the RCC and working towards her PhD thesis on 'Situating Climate Change in Chuuk: The cultural space of "nature" and "belonging" in Micronesia'.

Tobias Ide is a PhD candidate at the Research Group Climate Change and Security and the Institute of Geography, University of Hamburg, Germany. His research focuses on methodological and theoretical challenges in the research on climate change and conflict.

Gregory Kent is Senior lecturer in International Relations and Human Rights at the University of Roehampton, London.

Mark Levene has written extensively on genocide and anthropogenic climate change. His recent work includes *The Crisis of Genocide, The European Rimlands (1912–1953)*,

2 volumes (2013). He is founder of Rescue!History, and co-founder of Crisis Forum ('The Forum for the Study of Crisis in the 21st Century').

Andreas Pondorfer is PhD candidate and researcher at the Kiel Institute for the World Economy, Germany, funded by BIOACID (Biological Impacts of Ocean Acidification). His current research interests are African agricultural development in historical perspective, conflicts, climate change and risk.

Jürgen Scheffran is Professor at the Institute of Geography, Hamburg University, Germany, and head of the Research Group Climate Change and Security in the KlimaCampus Excellence Initiative. His field of research is on climate change, conflict and environmental migration; energy security; complex systems and human-environment interaction.

Janpeter Schilling is a Programme Officer in Climate Change and Security at International Alert in London and an Associated Researcher in the Research Group Climate Change and Security at Hamburg University, Germany. His field of specialisation is the linkages between climate change, vulnerability, adaptation and conflict.

Damien Short is Director of the Human Rights Consortium (HRC) and a Reader in Human Rights at the School of Advanced Study, University of London, UK. He is also the Director of the HRC's Extreme Energy Initiative, which researches the human rights impacts of unconventional fossil fuel extraction.

Jürgen Zimmerer is Professor of History at the University of Hamburg, Germany, and President of the International Network of Genocide Scholars. His research interests include German Colonialism, Comparative Genocide, Colonialism and the Holocaust, Critical Prevention Studies and Environmental Violence and Genocide. He is the author and editor of nine books and numerous articles.

Foreword

Jürgen Zimmerer[a]

[a]*Department of History, University of Hamburg/Cluster of Excellence (EXC177) 'Integrated Climate System Analysis and Prediction' (CliSAP), University of Hamburg, Germany*

In April 2014 the Intergovernmental Panel on Climate Change (IPCC) has yet again confirmed the reality of anthropogenic climate change and sketched some of its most dramatic consequences.[1] At the same time, the 20th anniversary of the beginning of the Rwandan genocide has focussed global attention on this gruesome act of collective violence. Many politicians and activists the world over have stated (yet again) their commitment to genocide prevention and thus renewed the call 'Never again' so enthusiastically proclaimed after World War II.[2]

Both events seem unconnected. Whereas research into the human consequences of climate change at least deals with violence, conflict and resilience, and therefore, one could argue also with genocide, genocide studies are almost completely ignoring climate change as a cause for violence, thus neglecting probably the biggest trigger of genocide in the twenty-first century. This ignorance comes at a huge price for genocide prevention and human security more generally.

This special issue aims at changing this and initiating a dialogue between scholars working in the areas of climate change and genocide. Research into genocide as well as climate change is a highly interdisciplinary endeavour, transcending the boundaries of established disciplines. The current special issue takes this into account by approaching the subject from a number of perspectives. Methodological and theoretical heterogeneity is intended and the various chapters should be read together.

In my opening article 'Climate Change, Environmental Violence and Genocide' I argue that anthropogenic climate change is the most fundamental challenge for humankind in the twenty-first century and will drastically alter the living conditions of millions of people, mainly in the global south. Therefore environmental violence, including resource crises such as peak fossil fuel, should be of major concern to genocide studies. The fact that this has not been the case is explained by the lopsided emphasis that traditional genocide scholarship places on ideology on the one hand and on legal prosecution on the other hand, leading to a much too narrow understanding of the driving forces of genocide. As a consequence of the latter I criticise the genocide prevention orthodoxy for its focus on humanitarian military intervention and argue for a new concept of sustainable prevention on the basis of global social justice.

Mark Levene and Daniel Conversi affirm in their article 'Subsistence Societies, Globalisation, Climate Change and Genocide: Discourses of Vulnerability and Resilience' the danger of structural genocide as a consequence of neoliberal globalisation and the nation-states that develop and operate alongside. They argue that 'business as usual' further endangers subsistence societies, which stand outside the market economy. On the one hand they identify those subsistence societies as particularly threatened by structural genocide; on the other hand they argue that it is the latter's resilience which could become a model for dealing with the consequences of globalisation and climate change more generally.

From a similar starting point Martin Crook and Damien Short recognise ecocide in their contribution 'Marx, Lemkin and the Genocide – Ecocide Nexus' as the most likely form and cause of genocide in the twenty-first century. To better understand those processes they argue for developing a theoretical apparatus that fuses political ecology and environmental sociology, particularly the concepts of Karl Marx, with the sociology of genocide inspired by Raphael Lemkin, the progenitor of the school of structural genocide. By way of illustration, the case of the First Nations in British Columbia and the ecocidal consequences of tar sands extraction is proffered.

Contemporary politics are most often performed either within the nation-state arena or an international system dominated by the former. Gregory Kent asks in his article 'Crystallisations of the Global Western State in the Era of Climate Change' how climate change and genocide will affect the latter. His analysis leads him to believe that the challenges posed by climate change will be a driving force for increased international cooperation.

Rebecca Hofmann offers in her article on 'Culturecide in Changing Micronesian Climates? About the Unintenionality of Climate Change' a case study on Micronesia, arguably one of the regions on the globe most affected by climate change and sea level rise. Since on the island of Chuuk, her research case, group identity is intrinsically linked to land, the threat that the island would need to be evacuated since it might be flooded in its entirety is much more than a threat to the material well-being of its inhabitants, but a fundamental danger to the group and its identity as such. Genocide studies will have to deal with cases like that, even if they are outside the remit of a traditional understanding of genocide.

The last two contributions deal with case studies from Africa. Andreas Exenberger and Andreas Pondorfer provide in their contribution 'Genocidal Risk and Climate Change: Africa in the Twenty-First Century' a meta-study of empirical research on the link between climate change and genocide and supplement their analysis of existing literature with an empirical analysis of their own on the relationship between climate change and food production and also the importance of institutional factors. From this base they identify states at risk, but also factors that could mitigate the outbreak of violence.

The lack of research on the link between climate change and violent conflict more generally is addressed by Jürgen Scheffran, Tobias Ide and Janpeter Schilling in their article on 'Violent Climate or Climate of Violence? Concepts and Relations with Focus on Kenya and Sudan'. The article offers a theoretical framework on climate change as a threat multiplier and offers an agent-based approach to understand the individual decisions for violence or cooperation in a given situation. Through in-depth analyses of conflicts in Darfur and Kenya they identify the reduction of socio-economic marginalisation, the development of resource-sharing mechanisms and the restriction of access to arms as a means to avoid entering spirals of violence.

As all contributions show, climate change is a major threat multiplier for violence or non-violent destruction and any understanding of prevention needs to take this into account. They offer a basis for much-needed critical prevention studies, which aim at sustainable prevention.

Notes

1. http://www.ipcc.ch
2. See for example: genocideprevention.be (accessed 13 April 2014).

Climate change, environmental violence and genocide

Jürgen Zimmerer

Department of History, University of Hamburg/Cluster of Excellence (EXC 177) 'Integrated Climate System Analysis and Prediction' (CliSAP), University of Hamburg, Germany

Anthropogenic climate change is the most fundamental challenge for humankind in the twenty-first century. Rising sea levels and the loss of agricultural land, severe weather changes and desertification are just some of the likely consequences that will drastically alter the living conditions of millions of people, mainly in the global south. Environmental violence, including resource crises such as peak fossil fuel, should be of major concern to genocide studies. This article argues that environmental violence is amongst the main driving forces of collective violence and that climate change will dramatically increase the likelihood of genocide occurring in areas at risk. On this basis, the article criticises genocide prevention orthodoxy for its focus on humanitarian military intervention and asks for a new concept of sustainable prevention on the basis of global social justice.

Each of the last three decades has been successively warmer at the Earth's surface than any preceding decade since 1850. In the northern hemisphere, 1983–2012 was likely the warmest 30-year period of the last 1400 years.[1]

Climate change

'Climate change' has had an astonishing career over the last three decades. Almost unknown in the 1980s – although the phenomena is much older – climate change had become somewhat of a hegemonic discourse at the end of the first decade of the new millennium. This 'career' is reflected in the history of the Intergovernmental Panel on Climate Change (IPCC), founded in 1988. It symbolises both the global nature of climate change and the necessity of addressing climate change in transnational cooperation. The IPCC's highest prominence, was marked by the year 2007, when it, along with former US Vice-President Al Gore, was awarded the Noble Peace Prize. As this article is being written, the IPCC has started presenting its Fifth Assessment Report, beginning as usual with the findings of Working Group 1 on the physical science basis of climate change; reports from Working Group 2 on 'Impacts, Adaptation and Vulnerability', Working Group 3 on 'Mitigation of Climate Change', and finally a 'Synthesis Report', will follow later in 2014.[2]

Despite all scientific evidence on climate change and its causes and consequences, it has become a matter of heated scientific and public debate. A series of 'scandals', such as 'climate gate', in which confidential emails of climatologists of the University of East Anglia were leaked, have been used to undermine the credibility of the scientific assessment, as has seemingly contradictory evidence, such as the apparent pause in the increase of temperature during the last 15 years, acknowledged now by the IPCC.[3] According to most scientists,[4] none of this fundamentally challenges the reality of climate change:

> Warming of the climate system is unequivocal, and since the 1950s, many of the observed changes are unprecedented over decades to millennia. The atmosphere and ocean have warmed, the amounts of snow and ice have diminished, sea level has risen, and the concentrations of greenhouse gases have increased[5]

Despite the overwhelmingly affirmative scientific evidence, these findings are met by fierce ideological opposition from certain politicians, lobbyists and journalists. The debate shows all the signs of the ideological rifts of past conflicts, with extreme representatives of both sides accusing each other of being denialists or alarmists. In a complete reversal of the actual position, those people arguing with scientific models and predictions are labelled as 'ideological', whereas the so-called climate sceptics, amongst whom are only a few serious scientists, claim objective truth for their position. They abuse, for example, qualified statements made by the IPCC, which tries to assess the likelihood of any given future event in percentages, as a lack of scientific evidence, whereas it could be seen with much more justification as proof of the scientific character of the entire endeavour. Often climate change critics claim to possess the absolute truth, an illusion which in itself is highly questionable and certainly unscientific.

Nevertheless, the close nexus between science and politics and the procedure where not only scientists but also diplomatic and other representatives of various governments negotiate at least the summary of the Assessment Reports has also drawn serious criticism from scientists involved in the work of the IPCC.[6]

One reason for this intense debate seems to be that the public – both supporters and critics of the IPCC – tend to concentrate on the apocalyptic dimension of climate change. And indeed, if some of the more extreme predictions come true, the world would become uninhabitable for humankind within a relatively short period of time. *The Guardian*, one of the most popular UK newspapers, terms this doomsday terminology 'Climate Calamity'.[7] However, it seems that this focus on the most extreme predictions, on the potentially apocalyptic consequences of unmitigated climate change, might in the end be to the detriment of the cause; in the light of the fierce battle regarding the apocalyptic dimension of passing tipping points and the danger of a runaway climate, the simple truth that climate change is already taking place and is already affecting the lives of millions of people seems to have been lost. It might be the tragedy of climate change research that the potentially truly apocalyptic character of climate change in the future obscures the view of the ground-level suffering occurring today. The fact that this suffering predominantly takes place – at least at this moment – in the global south might be a further explanation for this deficiency.

Nevertheless, in many ways the fierceness of this debate is testament to the 'success' of climate change as a concept, as it increasingly influences disciplines outside the narrow confines of climate sciences. In particular, human responses to climate change, both in preventing a further acceleration and in mitigating its effects, are studied in a variety of disciplines from economics[8] and political science[9] to anthropology,[10] from agriculture to engineering. Some scholars have even proclaimed the ascent of a new field of research: climate and culture, particularly focusing on the cultural and political consequences of climate change.[11] Since violence plays a role amongst the possible effects of climate change, conflict and peace research and security studies are also engaged with climate change research.[12] What is strangely amiss, despite the fact that many see in climate change the scourge of the twentieth century, and one would assume also of the twenty-first century, is the integration of genocide studies in the climate change debate. This article sketches some of the historical reasons for this lack and argues for the inclusion of climate change in genocide studies and vice versa. It does so with the aim of opening up genocide prevention to the challenges of environmental violence.

Climate change and environmental violence: a note on terminology

It appears that climate change in common usage has acquired a surplus of meaning that goes beyond the actual physical phenomenon or the set of physical phenomena originally subsumed under the term. Climate change has become a chiffre for man-made environmental change and its potentially catastrophic effects. Climate change in this broader sense, however, cannot be separated from the carbon economy and population growth, that is, the finiteness of resources. Both are not only amongst the main causes of climate change, but are themselves threat drivers as the unsatisfied need for resources and the increase in demand will, in all likelihood – as any resource scarcity – cause an increase in violence, conflict and war. To give just one example: fossil fuel is both a driving force of climate change and will, after its peak,[13] amplify environmental and other crises, as the lack of energy supply will need to be compensated for and a reduction in energy supply will most likely increase the likelihood of resource conflicts. At the same time, energy scarcity will limit the resources available to mitigate the consequences of climate change. This challenge will not change even if we factor in new 'extreme energy' techniques of oil and gas production that are expanding current fossil fuel reserves[14] as on the one hand the process of 'extreme energy' is itself unsustainable, while the greater extraction effort and risk involved also bring environmental problems of their own.[15]

Resource scarcity in general, however, is intensified by population growth as population growth results generally in an increase in consumption unless the unequal distribution of food, water, energy, etc. worsens. By the 1970s, the finite nature of resources had been identified and brought to the attention of a wider public. In their famous 1972 study, *The Limits to Growth*, Donella H. Meadows, Dennis L. Meadows, Jørgen Randers and William W. Behrens had modelled economic and population growth and come to the conclusion that the current lifestyle based on growing consumption was unsustainable.[16] Forty years later, the situation has worsened, exacerbated by – amongst other things – climate change. For the analysis of the societal impact and the prediction of future violence, climate change cannot be separated from general environmental change.[17]

Scientists have reacted to this challenge by introducing the concept of 'Anthropocene',[18] which as a geological epoch followed the 'Holocene' (a term that is used to describe the last 11,000 years), and began some 250 years ago with the advent of the carbon-based industrial revolution.[19] The concept of 'Anthropocene' acknowledges the interconnectedness of various human activities to an extent that, for the first time, humankind is seen as a primary mover of global physical phenomena. Rockström et al. have made use of this concept by attempting to identify and quantify thresholds by which various environmental changes became or will become irreversible and catastrophic. They have marked a set of nine 'planetary boundaries', the crossing of which would endanger the continental or global ecosystem:

(1) climate change;
(2) ocean acidification;
(3) stratospheric ozone depletion;
(4) interference with the global phosphorus and nitrogen cycles;
(5) rate of biodiversity loss;
(6) global freshwater use;
(7) land-system change;
(8) aerosol loading;
(9) chemical pollution.

Climate change took a prominent position:

> Anthropogenic pressures on the earth system have reached a scale where abrupt global environmental change can no longer be excluded. We propose a new approach to global sustainability in which we define planetary boundaries within which we expect that humanity can operate safely. Transgressing one or more planetary boundaries may be deleterious or even catastrophic due to the risk of crossing thresholds that will trigger non-linear, abrupt environmental change within continental- to planetary-scale systems.[20]

The idea to group together a multitude of factors, all of which have serious effects on the earth as a human habitat and tend to intensify each other, benefits also the research on the impact climate change has on human society. It draws attention to the fact that a multitude of factors, such as peak fossil fuels or population growth, all possess an environmental dimension. This being the case, the generic term 'environmental violence' seems useful. Violence in this understanding, however, is not only physical violence, let alone collective physical violence, but is also structural violence.[21] However, for the purpose of this article, I will concentrate on collective physical violence.

Amongst all forms of collective violence, none tends to attract greater public interest and to provoke greater moral outcry than genocide, the destruction of entire groups of people. The Herero and Nama in German south-west Africa (1904–1908), the Armenians, Assyrians and Pontic Greek in the Ottoman Empire (1915–1918), the Holocaust (1941–1945) or the genocide in Rwanda (1994) are the best known examples of genocide in the twentieth century. Even the relatively young twenty-first century has seen its share of genocidal violence in Darfur, the Democratic Republic of Congo and Kenya, to name just a few regions where elites have been accused of committing genocide.

Grass-roots movements engage worldwide in condemning genocide and work towards its prevention, mobilising thousands of people against it. They monitor hotspots around the world for signs of impending genocidal violence, try to establish early warning systems, and argue for the building up of international intervention capabilities. Although often reluctant to follow suit with real action, politicians regularly pledge their commitment to combating genocide and to the 'Never Again' pledge promised after the Holocaust. In 2011, the US even introduced their Atrocities Prevention Board, justifying it with national interests and moral obligation:

> Preventing mass atrocities and genocide is a core national security interest and a core moral responsibility of the United States.

> Our security is affected when masses of civilians are slaughtered, refugees flow across borders, and murderers wreak havoc on regional stability and livelihoods. America's reputation suffers, and our ability to bring about change is constrained, when we are perceived as idle in the face of mass atrocities and genocide. Unfortunately, history has taught us that our pursuit of a world where states do not systematically slaughter civilians will not come to fruition without concerted and coordinated effort.[22]

Neither climate change nor environmental violence feature prominently in these initiatives. The same is true for academic research on genocide, a burgeoning field with annual international conferences and several journals.

But how can scholars, activists and their organisations, which devote so much time and energy to the prediction of the future, ignore some of the biggest challenges for humankind? Why is it so difficult for the scholars of genocide to follow their colleagues in other disciplines in dealing with violence such as peace and conflict or security studies, where climate change has gained some prominence?[23]

In order to answer this question, a short summary of genocide both as a concept and as a field for academic research is necessary.

Genocide: intentionality, ideology and the international system

Historically, the concept of genocide is intrinsically linked to the Holocaust, to which in many ways it owes its very existence. Raphael Lemkin, a Polish-Jewish lawyer, who had to flee his native Poland after the German invasion coined the term in his seminal study on the German occupation policy in Eastern Europe *Axis Rule in Occupied Europe*. Here he defined genocide as

> a coordinated plan of different actions aiming at the destruction of essential foundations of the life of national groups, with the aim of annihilating the groups themselves. The objectives of such a plan would be disintegration of the political and social institutions, of culture, language, national feelings, religion, and the economic existence of national groups, and the destruction of the personal security, liberty, health, dignity, and even the lives of the individuals belonging to such groups. Genocide is directed against the national group as an entity, and the actions involved are directed against individuals, not in their individual capacity, but as members of the national group.[24]

After tireless campaigning by Lemkin and others the newly founded United Nations (UN) adopted the concept and made genocide a crime under international law. In their 1948 'Convention on the Prevention and Punishment of the Crime of Genocide' genocide was defined as 'any of the following acts committed with intent to destroy, in whole or in part, a national, ethnic, racial or religious group, as such:

(a) Killing members of the group;
(b) Causing serious bodily or mental harm to members of the group;
(c) Deliberately inflicting on the group conditions of life calculated to bring about its physical destruction in whole or in part;
(d) Imposing measures intended to prevent births within the group;
(e) Forcibly transferring children of the group to another group.[25]

The two central elements in this definition are the *intention* to destroy a specific, clearly defined *group* of people. Through political manoeuvring, political and social groups were excluded from the convention since the colonial powers of Great Britain and France were opposed to forcing the then-Soviet Union to recognise the latter. The murder of millions of Kulaks was not to be called genocide, nor should atrocities against postcolonial liberation groups be considered genocidal. The definition of groups has therefore from the beginning been a matter of controversy, particularly since neither 'national' nor 'ethnic' groups are clearly defined categories and 'race' is a highly contested term. There even seems to be an emerging consensus – at least amongst scholars – that there are no objective criteria for one's membership in a group, but that the victim group is rather arbitrarily defined by the perpetrators. It is part of the horror of genocide that as a victim one is categorised regardless of one's own perceptions or actions. The victims of genocide do not, thus, provoke the aggression meted out against them, neither can they choose not to belong to the victim group.

Closely linked to the question of the 'target' of genocide is that of intention, what Lemkin called a 'coordinated plan of different actions'. The UN convention was meant to be an instrument of international justice, even if it could not fulfil this task during the Cold War, and has left the international legal community wanting even in more recent times. Understanding genocide in a predominantly legal way with the focus on prosecution has led to a narrowing of the definition of 'intent'. Using the analogy of a murder case, where the prosecutor has to prove the *intention* to murder, the intention to commit genocide became the centre of attention. For example, did Milosevic order and intend the complete destruction of the Bosnian Muslims? Or did he 'just' aim at the ethnic cleansing of certain groups of people, in which some or many of them died as 'collateral damage'? What at a single murder trial would mean the difference between murder and manslaughter makes the identification of genocide extremely difficult. Even if prosecutors can prove that individual killings were intended, they still need to show that the entire group is the target. In 2005, for example, the report of the International Commission of Inquiry on Darfur concluded that whilst crimes against humanity and war crimes were committed in Darfur, which were perhaps no less heinous than genocide, still genocide could not be proven, because 'the crucial element of genocidal intent appears to be missing, at least as far as the central Government authorities are concerned'. What it meant by 'genocidal intent' can be extrapolated from the following excerpt:

> Generally speaking the policy of attacking, killing and forcibly displacing members of some tribes does not evince a specific intent to annihilate, in whole or in part, a group distinguished on racial, ethnic, national or religious grounds. Rather, it would seem that those who planned and organized attacks on villages pursued the intent to drive the victims from their homes, primarily for purposes of counter-insurgency warfare.[26]

This intention then had to be proven in court, and the best – and often the only proof – were written documents and orders. But they are difficult to come by, because genocidaires rarely put their murderous intentions down on paper, at least not in the form of precise orders. See, for example, the decade-long search for Hitler's order to exterminate the Jews of Europe. Historians have still not found the one single instruction that started the mass murder; however, there is no doubt about Hitler's responsibility for the Holocaust. Of course, in this case as in others, genocidal intention can be and is inferred from speeches and writings as well as from actions. Particularly, genocidal propaganda, which exists as a foundational factor in basically all genocides in a process of dehumanising 'the other' in order to ease their destruction, began to be seen as a much needed proof of intent.

This perspective was supported by the public perception of the Holocaust, which stood not only as the basis of the Genocide Convention but has in the meantime become a global chiffre for evil and for what harm humans can do to other humans.[27] Despite major advances in the historical sciences and Holocaust studies that draw a much more complex picture, there still seems to be a prevalent image of the Holocaust in which the Nazis from early on aimed at the physical murder of the Jews, only claimed power, and started World War II to solve the 'Jewish question', i.e. to murder the Jews of Europe (and possibly beyond). The public narrowing of the focus on Hitler and German anti-Semitism has deeply entrenched our perception of genocide as a crime primarily, if not exclusively, based on ideology. It appears as a deeply irrational act juxtaposed against seemingly rational forms of violence, such as violence in self-defence and warfare. This perception is enshrined in the focus on the ideological elite, i.e. the Nazis with their occult rituals and their Aryan (= racial) fantasies, contrasted against the bureaucratic and military elite often portrayed as un-ideological and un-political. Furthermore, the fact that the Holocaust was carried out by one of the most efficient bureaucratic regimes in history emphasises a perspective that state elites and the state apparatus commit genocide. For many, the ideological elite (Nazis) hijacked an entire state (Germany) mainly to carry out its crime based on a racial dystopia.[28]

This model of genocide has serious consequences for understanding genocide and also the means by which to prevent it. Combined with an often far too simplistic understanding of ideology, it leads to a narrowing of prevention to a limited number of trigger points, which allegedly serve as detectors for an early warning system.[29] The concentration on ideological causes seems to have led to ignoring systemic causes. As a consequence, prevention is reduced to intervention.[30] To discuss only one, albeit quite prominent, example: Gregory Stanton has developed the highly influential concept of the 'Eight Stages of Genocide', aimed at identifying genocides before they happen in order to prevent them and serving as basis for other attempts to establish early warning systems. Genocide, according to Stanton,

> is a process that develops in eight stages that are predictable but not inexorable. At each stage, preventive measures can stop it. The process is not linear. Logically, later stages must be preceded by earlier stages. But all stages continue to operate throughout the process.[31]

Stanton then identifies in his concept, which he first developed as a briefing paper for the US State Department, the following stages:

- Classification
- Symbolisation
- Dehumanisation
- Organisation
- Polarisation
- Preparation
- Extermination
- Denial

If they are to be detected in regimes, then genocide has to be seen as possible and immanent. To prevent genocide is to stop classification, to stop hate propaganda and, in stage seven, ultimately to intervene in order to stop the ongoing extermination.[32] Leaving aside the problem of military intervention, which cannot be discussed here for reasons of space,[33] the problem with this approach lies in the fact that it describes a crime seemingly without a cause. Stanton explains the classification, from which all other stages derive, as follows:

> All cultures have categories to distinguish people into 'us and them' by ethnicity, race, religion, or nationality: German and Jew, Hutu and Tutsi. Bipolar societies that lack mixed categories, such as Rwanda and Burundi, are the most likely to have genocide.[34]

This is problematic in several ways. First, the statement that some societies 'lack mixed categories' seems to accept categories as being real, whereas they are constructed. Second, it can be argued that racism is particularly violent in societies where the boundary between the 'races' is blurred or under threat, where 'mixed groups' do exist.[35] In other words, ideology creates a difference, where the sociological reality tells a different story. See, for example, the obsession of the Nazis to define who was a Jew and who was not. The ideological rigor was partly due to the fact that Jews where widely integrated in German society and not easily detectable. However, the most serious problem in Stanton's theory is that he describes a situation but fails to address the most important question: why and under which conditions do the binary constructions of society lead to genocide? Binary constructions seem to be necessary, but not sufficient. If all cultures have an 'us and them' dichotomy, then it explains nothing about why in some cultures genocides are committed and in others they are not. In other words, what are completely missing are the causes for genocide, only the tools are described for carrying it out.

But also, why people would start classifying each other or certain groups, marking each other out and dehumanising each other, is not even elaborated in Stanton's theory. The concept, valuable as it may be for identifying an impending or ongoing genocide, does not aid our understanding of the causes for genocide. Why genocides take place is not explained; the space for the causes is left blank. This blank however is in the common perception filled with references either to traditional 'ethnic' stereotypes such as African atavistic violence or with recourse to ideology or irrational beliefs. In many ways it is an ahistorical approach to genocide and its prevention and it ignores the deeper roots of genocide. Models like this remain on the surface and construct genocide as a mainly ideologically motivated regime of criminals and psychopaths, who, driven by an irrational hatred, have hijacked a state and abuse this powerbase to carry out their irrational murderous ideology.

Furthermore genocidal regimes are seen not only as criminal and irrational, but they are also characterised as aberrations from the proper functioning of modern regimes, as signs of rogue or failed states in an otherwise well-functioning and rational international political system. This image of genocide as a consequence of the dysfunction of the state, which is seen as the prime agent of the international system, resonates also with the much heralded UN initiative of a 'Responsibility to Protect' (R2P):

> 138. Each individual State has the responsibility to protect its populations from genocide, war crimes, ethnic cleansing and crimes against humanity. This responsibility entails the prevention of such crimes, including their incitement, through appropriate and necessary means.[36]

This perspective on the international system has consequences. If states and their governments do not fulfil this task of protecting their population, then the international community is legitimised and even requested to interfere, even to intervene.[37]

However labelling genocidal regimes as 'rogue' or 'failed' states creates a 'dark other' to the Western/northern state, allowing the international community to respond to their 'misbehaving' in the modus of police action,[38] that is with short, precise, violent action, and, at the same time, stabilises the international system. Singling out the 'rogue' regimes attests to the innocence of the rest. Emphasising the irrational nature of genocidal violence affirms the rationality of all other actors in the global system. Under the guise of humanitarianism, we find logic at work, with a clear perpetrator and victim dichotomy and identifying clear culprits who need to be brought to justice. Consequently, the suggested course of action is that of stopping the symptoms instead of addressing the causes. There is no space – and within this logic of prevention no need – for analysing and changing the systemic, transnational and global roots of genocide.

Genocide, environmental violence and rational choice

During the last decade, however, research into global historical cases of genocide has started to produce a different picture of genocide. Ideology has been decentred, intention redefined and our understanding of genocide widened in this process.[39] Genocide can, therefore, be seen as a crime aimed at a social collective defined by the perpetrator. It is still intrinsically linked to the problem of identity, since both the victim and the perpetrator group are constructed in a reciprocal process of othering in which hybridity, ambiguity and multiple identities are substituted by unambiguousness identities: contact and communication by separation, and communication by expulsion or murder.

Expulsion is here included as a form of genocide since it most often leads to the destruction – physical or cultural – of the expelled group, and can thus be seen as genocidal. So far, research has focused little, if at all, on the ways these identities were constructed (racial theory, etc.), the reasons why societies resort to creating absolute dichotomies that in turn serve as precursors to violence, or why traditional dichotomies are suddenly framed in a way that resorting to violence seems likely, desirable, or even inevitable. This is where environmental violence and globalisation comes into play.

Historically the competition for territory is a primary cause of violence. Space is a finite quantity for which people (by definition indefinite in their numbers) compete. The need for land can be real or imagined (it can include imaginative landscapes, for example, plans for settlements, economic or agricultural use, or fear of land shortage). Early globalisation in the form of European colonialism affected this in three ways: The movement of people (settlement), the inclusion of distant regions in the emerging world economy (agriculture, mining, hunting, trading with certain impacts on people living there) and an increase in communication over vast distances (exchange of personnel, representations, learning experiences).

Globalisation and its impact on the history of genocide has only fairly recently started to attract scholarly attention. The best analysis so far can be found within the research into genocides in the colonial context.[40] As I have argued elsewhere, colonialism, in particular settler colonialism, can be seen as the control of space (land) on the basis of race.[41] It is – if nothing else – land grabbing by the colonisers on a truly global scale. Genocidal violence accompanied the colonial settlement process, particularly in settler colonies. Although land was not really scarce, since many regions particularly in North America and Australia were not overly densely populated, there was no 'terra nullius', no empty land, either.

Soon, competition for land began between indigenous populations and newcomers. This competition was fuelled by a change in land use patterns introduced by the settlers to start producing for the colonial and emerging global market. The large-scale introduction of sheep grazing is a good example.[42] To turn sheep farming into a global export, large areas populated by indigenous peoples needed to be cleared. This most often occurred through local initiatives and militias, sometimes even opposed by the central colonial government.[43] Ideology in the form of racism and ideas of European and Christian superiority supported this by allowing the newcomers to place the local population outside the sphere of moral responsibility; an indispensable precondition of genocide. Victims need to be excluded from the group of those whom 'we are required to protect and to take seriously, and opposite which we have responsibilities', Helen Fein clarifies.[44] In the colonial context, this means that moral obligation only covered the European colonisers; moral constraints in behaviour were only valid within this group.

However, ideology was not the prime mover, but was instead merely an enabling factor. The cause can rather be seen in the economic decision to produce for a larger colonial or even global market. Economic interests were at the forefront when it came to expelling and destroying entire peoples. These economic decisions are commonly described as 'rational'; in this case, an economic decision was a rational decision for genocide. Genocide in this sense is violent communication between two groups of people in competition for space and, in general, for more resources.

In the colonial context, the resource in question was mainly land. Interestingly, land was also foundational for Raphael Lemkin's understanding of genocide based on his analysis of the Nazi occupation policy:

> Genocide has two phases: one, destruction of the national pattern of the oppressed group; the other, the imposition of the national pattern of the oppressor. This imposition, in turn, may be made upon the oppressed population which is allowed to remain or upon the territory alone, after removal of the population and the colonization by the oppressor's own nationals.[45]

Lemkin's concept not only offers categories to link distinct instances of genocide in world history – in what I have called in a different context 'an archaeology of genocide'[46] – but also argues for the inclusion of cultural destruction as a form of genocide. That genocide is not necessarily or not primarily about mass killing, but about group destruction is a point often overlooked. This destruction can also be seen in the destruction of the cultural foundation of group identities and can also be brought about by a change in land use patterns, i.e. the competition for resources.[47]

Genocide as a resource-led form of violence is not limited to classical colonial situations, as new readings of the German genocides during World War II prove. In the last two decades, a number of studies have simultaneously widened and deepened our understanding of Nazi genocidal violence, the underlying motifs and its victims.[48] By putting the Holocaust and other crimes into the context of the actual political and military situation and assessing developments on the ground, scholars have deconstructed the simplistic notion of central planning and clear-cut, top-down responsibility for the genocides based on a simplistic ideology of racial hatred, important as the latter, particularly in the case of the Jews, certainly has been. By including local developments on and behind the Eastern front this has directed attention to more general ideas of Germany's Eastern Empire.[49] *Lebensraum* ('living space') in the East was seen as the necessary answer to a perceived shortage of space due to overpopulation, a fear quite prominent in European thinking of the nineteenth and twentieth centuries. Although the Malthusian prediction of overpopulation did not materialise, thinking along those lines did influence German policy; the fear of resource scarcity (at the heart of Malthus' prediction) and the social Darwinian understanding of nations as being bio-political entities caught in a permanent struggle against each other and in which only the 'survival of the fittest' seemed possible led to the dangerous conclusion that this bio-political body needed to be strengthened by cleansing the 'body' from so-called 'parasites' (in the Nazi ideology: Jews, Romani people, homosexuals, the handicapped, etc.), bringing all 'true ethnic' Germans back into the nation and by making space available for the 'healthy' members and their (numerous) offspring. What many see as a monolithic Nazi genocide was, in reality, quite a complex interplay of cleansing (and eventually killing) the unwanted and concentrating and resettling the wanted. Many of the decisions to murder the Jews in certain areas were made locally in order to solve a regional 'problem' of overpopulation due to the influx of German settlers. It seemed to them to be a rational solution to a real problem. Of course the 'problem' was self-created by the German administration that developed resettlement plans on the basis of an imagined shortage of land. Land also played a key role for Nazi planning regarding the fate of the Slavs. In order to clear the Western parts of the Soviet Union, the SS developed concepts to move and murder as many as 80 million Slavs. According to the so-called *Generalplan Ost* (General Ordinance East) most of the population of the Soviet Union west of the Ural Mountains were to be expelled to Siberia. Knowing full well that this eastern province would never be able to feed so many people, their starvation and eventual death was willingly accepted by German planners. Again, gaining land for German settlement was the main objective. Due to the changing fortunes of war, the plan could never be put fully into practice. However, given the importance of German settlement fantasies and the population movements already initiated, local decisions to solve local supply 'problems' by mass murder could appear rational, albeit deeply inhuman.

It goes against the grain of human self-conception to call the decision for mass murder a rational decision and therefore public images of mass violence often resort to irrationality as an explanation.[50] However, the complex relationship between rationality and irrationality is not entirely new. Already Theodor Adorno and Max Horkheimer have emphasised the 'Dialectics of Enlightenment'[51] and Zygmunt Baumann[52] has shaped the image of the modern state as the gardening state, i.e. the gardener who decides on a rational basis which plants are allowed to grow and which need to be weeded out. Nazi policies also demonstrate that rational decisions can be based on irrational ideas, world-views and fears.

In general, the relationship between the rational and the ideological in genocide is quite complex. Any perceived threat of resource scarcity, etc., might lead to the fear that there is not enough water, food, grain, oil, land, and so on, and, therefore, some people will have to be excluded from its use. This population, perceived as 'superfluous', would have to disappear, leave the region, or be killed. It is this interplay between ideology and rational choice, between crises (real or imagined) and the decision for mass destruction that opens up genocide studies to discussions on environmental violence. If we assume that environmental violence will lead to systemic stress and to multiple crises, then we are to expect an increase in genocidal violence.

Exclusions, however, do not happen indiscriminately. Here, ideology, traditions of inclusion and exclusion, and histories of violence come into play. If certain people or groups of people are targeted, traditional antagonisms and dichotomies, the binary opposites in many societies between several groups (according to cultural, religious or regional differences), are exploited. Once the logic has been accepted that there are not enough resources for everybody, targeting one's traditional enemies might seem to make sense.

Nonetheless, if we accept that resource scarcity can create genocidal violence then we should alter our understanding of the role of ideology, of intention, and ultimately of prevention. Ideology will still be important; however, it might not be the initial cause of violence in each and every case. Rather, resource scarcity – real or perceived – could serve as a cause as well as part of an ideology, whereby ideology becomes the means by which allegedly superfluous human beings are identified.

Genocide, environmental change and sustainable prevention

Obviously, theories of genocide prevention based on an understanding of genocide as a purely ideological crime and resting on theoretical models with various stages to identify potential genocidal hotspots are inadequate.[53] And so ideas of prevention are still following a control paradigm – seeing instances of genocide as limited, distinct events that can be controlled by international 'police action'. Military intervention in particular does not seem an appropriate tool to achieve sustainable prevention. If genocidal crises erupt in greater numbers – which might be an effect of the stress caused by environmental change on individual societies as well as the global system in its entirety – the international community will simply not be able to cope through sanctions and military interventions. Furthermore, there is a limit to the scale, intensity and duration of military force that can be deployed before the intended consequences (stopping violence) are outweighed by adverse effects, such as becoming partisan in the conflict, fuelling anti-Western sentiments, inflicting casualties on innocents, and suffering casualties, which makes support at home for the mission crumble. And third, containing the conflicts in the regions of their outbreak might prove increasingly difficult. Violence tends to spread to neighbouring countries, be it as the result of strategies of escalation or in the form of refugees. With presumably ever larger regions being affected, more and more people will try to reach the lesser affected regions, many of them in the global north. Refugee streams on a large scale cannot be stopped forever by the richer nations of the global north without undermining the moral foundations of their actions, if preventing refugees from reaching the wealthy states of 'the West' is, indeed, morally justifiable at all. Therefore, sustainable prevention cannot be achieved under the control paradigm; a radical rethink is necessary.

What is required is sustainable prevention based on a 'global social justice approach'. By that, I mean an integral approach to the problem of genocide that accepts the social dimension of the issue. Rudolph Joseph Rummel has argued that liberal democracies are less prone to commit genocidal actions than other states.[54] Putting aside the question of whether this statement comes at the expense of ignoring settler colonial violence, which was partly set well within emerging democratic structures, there is yet another way to read the historical evidence: liberal democracies are, historically, rather wealthy and stable societies. Using Rummel as an analogy, we could conclude that richer societies are less prone to genocidal violence than poorer ones or, phrased differently, societies in crises are more prone to resort to genocidal violence than others.[55] A social justice approach therefore understands global social and economic inequality as a major root cause of violence, since the latter destabilises social communities and increases the likelihood of war over resources, of intra-societal fights for wealth distribution, and of an export of violence through refugees, to name just a few examples.

If we accept, however, extreme social inequality and social tensions as one of the root causes of genocide, effectively addressing global social injustice would be a means – if perhaps the most promising one – of preventing genocide. That would mean taking on a serious social problem of almost unimaginable proportions, one that cannot be solved by sending a few soldiers to fight a proxy war against genocide on behalf of the saturated majority at home. And it would mean that in contrast to lobbying for regime change 'over there' in the global south, we would also have to fight for changes in the economic consumer system 'over here' in the global north. All the more so in the light of climate change, because global injustice is getting worse with its effects.

This also has serious consequences for the perpetrator-victim position. No longer would scholars and genocide prevention activists from the global north be – by virtue of their location alone – the 'good guys' and the perpetrators always in the global south, but light would be shed on the very uncomfortable fact that genocide scholars and others are perpetrators as well. Whoever is serious about prevention will have to address the question of the current global hegemonic system, which perpetuates the highly unjust distribution of wealth on our planet, a problem that will get even worse due to climate change and the scarcity of many vital resources (e.g. fossil fuels, potable water); an ideology of mere economic growth, claiming that the poorer regions of the planet will eventually be pulled out of poverty, is no longer in itself a viable or sensible option. On the contrary, cutting back on consumption is necessary purely in order to save the climate, without any element of wealth transfer between richer and poorer nations.[56]

Thus, in order to achieve sustainable prevention we have to address the wider issues of global injustice, its impact on societies, and its potential for igniting violence. For activists and lobbyists campaigning to prevent genocide, this means also addressing questions about the current world order.

Insurmountable as this task seems, genocide studies cannot shy away from what is probably the biggest challenge of the twenty-first century. This special issue on climate change and genocide will make a start by exploring the connection between climate change and genocide from a variety of disciplinary and geographical perspectives. I am delighted that natural and political scientists, anthropologists, economists and historians were prepared to join this experiment. This is certainly not the last word on the issue. However, if this volume initiates a change in how we think about the future of collective violence and prevention by accepting that environmental violence is a key part of the history of genocide, if this volume initiates discussion on the global nature of the problem in the sense that we might all become victims and many of us are perpetrators, the special issue will have fulfilled its main tasks. The other is a plea for true interdisciplinarity. Since the problems are highly complex, so need to be the attempts to understand them.

Acknowledgements

The publication of this special issue of the *International Journal of Human Rights* was supported by the Cluster of Excellence 'Integrated Climate System Analysis and Prediction – CliSAP' (EXC177), University of Hamburg, funded by the German Science Foundation (DFG). I would like to thank Nils Schliehe for his help with the formal aspects of the editing, the co-editor of the *IJHR*, Dr Damien Short, and all reviewers for their comments and particularly Dr Elisabeth Hope Murray, both for her help with the language and her valuable criticism particularly on the issues of ideology, where she markedly differs from my approach.

Notes

1. IPCC WGI AR5 SPM, 27 September 2013, 2.
2. http://www.ipcc.ch.
3. The IPCC Summary Report concedes this pause, but sees no fundamental challenge to its findings.
4. John Cook, Dana Nuccitelli, Sarah A Green, Mark Richardson, Bärbel Winkler, Rob Painting, Robert Way, Peter Jacobs and Andrew Skuce, 'Quantifying the Consensus on Anthropogenic Global Warming in the Scientific Literature', *Environment Research Letters* 8, no. 22013. http://iopscience.iop.org/1748-9326/8/2/024024/article (accessed 27 March 2014).

5. IPCC WGI AR5 SPM, 27 September 2013, 2.
6. Hans von Storch for example has criticised the politicisation of the Climate Change (CC) and even made it a focus of scholarly analysis. Hans von Storch and Werner Kraus, *Die Klimafalle: Die gefährliche Nähe von Politik und Klimaforschung* (München: Hanser, 2013).
7. 'IPCC: 30 Years to Climate Calamity if We Carry on Blowing the Carbon Budget', *The Guardian* 27 September 2013. http://www.theguardian.com/environment/2013/sep/27/ipcc-world-dangerous-climate-change (accessed 18 December 2013).
8. See Andreas Exenberger and Andreas Pondorfer, 'Genocidal Risk and Climate Change: Africa in the Twenty-First Century', *The International Journal of Human Rights* 18, no. 3 (2014): 350–68.
9. See Gregory Kent, 'Crystallisations of the Global Western State in the Era of Climate Change', and Martin Crook and Damien Short, 'Marx, Lemkin and the Genocide – Ecocide Nexus', *The International Journal of Human Rights* 18, no. 3 (2014): 320–35 and 298–319.
10. See Rebecca Hofmann, 'Culturecide in Changing Micronesian Climates? About the Unintenionality of Climate Change', *The International Journal of Human Rights* 18, no. 3 (2014): 336–49.
11. Claus Leggewie and Harald Welzer, *Das Ende der Welt wie wir sie kannten* (Frankfurt/M: Fischer, 2010). See, for a similar approach, Mark Levene, 'Climate Blues: Or How Awareness of the Human End Might Re-instil Ethical Purpose to the Writing of History', *Environmental Humanities*, no. 2 (2013): 153–73.
12. See Jürgen Scheffran, Tobias Ide and Janpeter Schilling, 'Violent Climate or Climate of Violence? Concepts and Relations with Focus on Kenya and Sudan', *The International Journal of Human Rights* 18, no. 3 (2014): 369–90.
13. The International Energy Agency argues that we have already reached the peak of conventional fossil fuel. See International Energy Agency, *World Energy Outlook, 2013 Factsheet*, 3. http://www.worldenergyoutlook.org/media/weowebsite/factsheets/WEO2013_Factsheets.pdf (accessed 27 March 2014).
14. The German intelligence agency BND has forecasted that peak oil is not happening in the foreseeable future due to new production techniques such as fracking. 'BND warnt vor Klimawandel Konflikten', *Der Spiegel*, 1 November 2013. http://www.spiegel.de/wissenschaft/natur/geheimdienst-analyse-bnd-warnt-vor-klimawandel-konflikten-a-931290.html (accessed 5 November 2013).
15. See for example http://extremeenergy.org/2013/07/25/defining-extreme-energy-a-process-not-a-category/ (accessed 27 March 2014).
16. Donella H. Meadows, Dennis L. Meadows, Jørgen Randers, and William W. Behrens, *The Limits to Growth* (New York: Universe Books, 1972).
17. For an assessment of the original study see: Graham M. Turner. 'A Comparison of the Limits to Growth with 30 Years of Reality', *Global Environmental Change* 18 (2008): 397–411. http://www.csiro.au/~/media/CSIROau/Divisions/CSIRO%20Sustainable%20Ecosystems/SEEDPaper19_CSE_publication%20Standard.pdf (accessed 27 March 2014).
18. See also Mark Levene and Daniel Conversi, 'Subsistence Societies, Globalisation, Climate Change and Genocide: Discourses of Vulnerability and Resilience', *The International Journal of Human Rights* 18, no. 3 (2014): 281–97.
19. P.J. Crutzen, 'Geology of Mankind: The Anthropocene', *Nature*, no. 415 (2002): 23; W. Steffen, P.J. Crutzen, and J.R. McNeill, 'The Anthropocene: Are Humans Now Overwhelming the Great Forces of Nature?', *Ambio*, no. 36 (2007): 614–21.
20. J. Rockström, W. Steffen, K. Noone, Å. Persson, F.S. Chapin, III, E. Lambin, T.M. Lenton, M. Scheffer, C. Folke, H. Schellnhuber, B. Nykvist, C.A. De Wit, T. Hughes, S. van der Leeuw, H. Rodhe, S. Sörlin, P.K. Snyder, R. Costanza, U. Svedin, M. Falkenmark, L. Karlberg, R. W. Corell, V.J. Fabry, J. Hansen, B. Walker, D. Liverman, K. Richardson, P. Crutzen, and J. Foley, 'Planetary Boundaries: Exploring the Safe Operating Space for Humanity', *Ecology and Society* 2, no. 14 (2009): 32. http://www.ecologyandsociety.org/vol14/iss2/art32/.
21. Johan Galtung, 'Violence, Peace, and Peace Research', *Journal of Peace Research* 3, no. 6 (1969): 167–91.
22. Barack Obama, 'Presidential Study Directive/PSD-10′, Press Release, 4 August 2011. http://www.whitehouse.gov/the-press-office/2011/08/04/presidential-study-directive-mass-atrocities (accessed 1 November 2013).

23. Jürgen Scheffran, Michael Brzoska, Hans Günter Brauch, Peter Michael Link and Janpeter Schilling, ed., *Climate Change, Human Security and Violent Conflict: Challenges for Societal Stability* (Berlin: Springer, 2012).
24. Raphael Lemkin, *Axis Rule in Occupied Europe* (Washington, DC: Carnegie Endowment for International Peace, Divisions of International Law, 1944), 79. For new approaches to Lemkin see Donna-Lee Frietze, ed., *New Approaches to Raphael Lemkin*, Special Issue of the *Journal of Genocide Research* 3, no. 15 (2013); John Cooper: *Raphael Lemkin and the Struggle for the Genocide Convention* (London: Palgrave Macmillan, 2008).
25. Convention on the Prevention and Punishment of the Crime of Genocide. http://www.unhchr.ch/; html/menu3/b/p_genoci.htm (accessed 5 November 2013).
26. Report of the International Commission of Inquiry on Darfur. http://www.un.org/News/dh/sudan/com_inq_darfur.pdf (accessed 17 October 2013).
27. See for example the study by Daniel Levy and Natan Sznaider, *The Holocaust and Memory in the Global Age* (Philadelphia, PA: Temple Press, 2005).
28. Christian Gerlach has quite rightly criticised the overall focus in genocide studies on governmental elites, ideology and a single perpetrator, single victim dichotomy as well as the idea of perpetrators acting according to established blueprints. Christian Gerlach, *Extreme Violent Societies. Mass Violence in the Twentieth-Century World* (Cambridge: Cambridge University Press, 2010).
29. For a far more advanced and complex model see the 'Minority at Risk Project'. http://www.cidcm.umd.edu/mar/ (accessed 12 March 2014).
30. I have touched upon this subject elsewhere: See Jürgen Zimmerer, 'Beyond Gaddafi: Sustainable Prevention in the Face of Environmental Injustice', *Journal of Genocide Research* 1–2, no. 13 (2011): v–vii.
31. Gregory Stanton, 'The 8 Stages of Genocide'. http://www.genocidewatch.org/genocide/8stagesofgenocide.html (accessed 14 October 2013).
32. Ibid. See the impressum for the reference to the State Department.
33. For a critical assessment of humanitarian interventions see: Alex J. Bellamy, 'Military Intervention', in *Oxford Handbook of Genocide Studies*, ed. D. Bloxham and A.D. Moses (Oxford, Oxford University Press, 2010), 597–616. For a critical assessment of the potential abuse of humanitarian intervention by the US with particular reference to the R2P, see Daniel Feierstein, 'The Good, the Bad, and the Invisible. A Critical Look at the MARO Report', *Genocide Studies and Prevention* 6, no. 1 (2011): 39–44; and Daniel Feierstein, 'Getting Things into Perspective', *Genocide Studies and Prevention* 4, no. 2 (2009): 155–60.
34. Stanton, 'The 8 Stages of Genocide'.
35. For a colonial example see, Jürgen Zimmerer, 'Deutscher Rassenstaat in Africa. Ordnung, Entwicklung und Segregation in "Deutsch-Südwestafrika" (1884–1915)', in *Gesetzliches Unrecht. Rassistisches Recht im 20. Jahrhundert* [=Jahrbuch zur Geschichte und Wirkung des Holocaust 2005] (Frankfurt/M.: Campus, 2005), 135–53.
36. United Nations, General Assembly 2005 World Summit Outcome Document. http://www.un.org/en/preventgenocide/adviser/pdf/World%20Summit%20Outcome%20Document.pdf#page=30 (accessed 6 November 2013).
37. For genocide and international relations see also: Adrian Gallagher, *Genocide and its Threat to Contemporary International Order* (Houndmills: Palgrave, 2013); and Martin Shaw, *Genocide and International Relations, Changing Patterns in the Transitions of the Late Modern World* (Cambridge: Cambridge University Press, 2013).
38. For the concept of 'police action' see: Paul Rogers, *Losing Control: Global Security in the Twenty-First Century* (London: Pluto Press, 2010).
39. I have touched upon this in: Jürgen Zimmerer, 'From the Editors: Environmental Genocide? Climate Change, Mass Violence and the Question of Ideology', *Journal of Genocide Research* 3, no. 9 (2007): 349–51. For an overview of various cases of genocide see: Donald Bloxham and A. Dirk Moses, eds, *The Oxford Handbook of Genocide Studies* (Oxford, Oxford University Press, 2010).
40. See for a general overviews: Mark Levene, *Genocide in the Age of the Nation State*, Volume 2 (London: I.B. Tauris, 2005); A. Dirk Moses, ed., *Genocide and Settler Society: Frontier Violence and Stolen Indigenous Children in Australian History* (New York: Berghahn Books, 2004); A. Dirk Moses, ed., *Empire, Colony. Conquest, Occupation, and Subaltern Resistance in World History* (New York: Berghahn Books, 2008); A. Dirk Moses and Dan Stone, ed., *Colonialism and Genocide* (London: Routledge, 2007).

41. Jürgen Zimmerer, 'Colonialism and the Holocaust. Towards an Archaeology of Genocide', in *Genocide and Settler Society: Frontier Violence and Stolen Indigenous Children in Australian History*, ed. A. Dirk Moses (New York: Berghahn Books, 2004), 49–76; Jürgen Zimmerer, 'The Birth of the "Ostland" Out of the Spirit of Colonialism: A Postcolonial Perspective on the Nazi Policy of Conquest and Extermination', *Pattern of Prejudice*, no. 39 (February 2005): 197–219; Jürgen Zimmerer, *Von Windhuk nach Auschwitz. Beiträge zum Verhältnis von Kolonialismus und Holocaust* (Münster: LIT-Verlag 2011) (An English edition will be published with Routledge in 2014).
42. Norbert Finzsch, '"[. . .] Extirpate or Remove that Vermine": Genocide, Biological Warfare, and Settler Imperialism in the Eighteenth and Early Nineteenth Century', *Journal of Genocide Research* 2, no. 10 (2008): 215–32; Norbert Finzsch, 'Pre-Frontier, Landnahme und Sozioökologische Systeme in Australien, 1788 bis 1901', in *Themenportal Europäische Geschichte* (2013). http://www.europa.clio-online.de/2013/Article=614; Mohamed Adhikari, *The Anatomy of a South African Genocide: The Extermination of the Cape San Peoples* (Cape Town: University of Cape Town Press; and Athens, OH: Ohio University Press, 2010).
43. See Zimmerer, 'Colonialism and the Holocaust'.
44. Helen Fein, 'Definition and Discontent: Labelling, Detecting, and Explaining Genocide in the Twentieth Century', in *Genozid in der modernen Geschichte*, ed. Stig Förster and Gerhard Hirschfeld (Münster: LIT-Verlag, 1997), 11–21, 20.
45. Lemkin, *Axis Rule in Occupied Europe*, 79. Lemkin himself did not see the concept of genocide as being limited to the Third Reich and its crimes, but also based it on his understanding of the Armenian genocide. Later he developed a global concept and worked on a three-volume global history of genocide, unfortunately unfinished and unpublished, in which he intended to devote a great deal of attention to genocide in the colonial context. For Lemkin as a historian of mass violence, see Dominik J. Schaller and Jürgen Zimmerer, ed., *The Origins of Genocide. Raphael Lemkin as a Historian of Mass Violence* (London: Routledge, 2009).
46. Zimmerer, 'Colonialism and the Holocaust'.
47. For an introduction to Lemkin's thinking on that matter see Dirk Moses and Raphael Lemkin, 'Culture, and the Concept of Genocide', in *The Oxford Handbook of Genocide Studies*, ed. Donald Bloxham and A. Dirk Moses (Oxford: Oxford University Press, 2010), 20–41.
48. Most of the scholars responsible for this turn in the history of the Third Reich have presented their findings in Ulrich Herbert, ed., *National Socialist Extermination Policies: Contemporary German Perspectives and Controversies* (New York: Berghahn, 2000). See there for further references.
49. For the idea of reading the German war against the Soviet Union as an attempt at empire building, see Zimmerer, *Windhuk nach Auschwitz*. Since the beginning of this discussion, overviews have been presented by Mark Mazower, *Hitler's Empire. Nazi Rule in Occupied Europe* (London: Penguin, 2008); and Shelley Baranowski, *Nazi Empire: German Colonialism and Imperialism from Bismarck to Hitler* (Cambridge: Cambridge University Press, 2010).
50. Charles Anderton, 'Choosing Genocide. Economic Perspectives on the Disturbing Rationality of Race Murder', *Defence and Peace Economy* 5–6, no. 21 (2010): 459–86.
51. Max Horkheimer and Theodor Adorno, *The Dialectics of Enlightenment* (Stanford, CA: Stanford University Press, 2002 [1944]).
52. Zygmunt Bauman, *Modernity and the Holocaust* (Ithaca, NY: Cornell University Press, 1989); Zygmunt Bauman, *Modernity and Ambivalence* (Cambridge: Polity, 1991).
53. I have sketched out this idea briefly in Jürgen Zimmerer, 'Beyond Gaddafi: Sustainable Prevention in the Face of Environmental Injustice: Some Words of Praise but Also Caution!', *Journal of Genocide Research* 3, no. 13 (2011): 209–15; and Zimmerer, 'Genocide Studies for the Twenty-first Century: A Departing Editor's Perspective', *Journal of Genocide Research* 3, no. 13 (2011): 205–8.
54. See for example Rudolph Rummel, *Death by Government: Genocide and Mass Murder in the Twentieth Century* (New Jersey: Transaction Publishers, 1994); and Rudolph Rummel, *Statistics on Democide. Center on National Security and Law* (Charlottesville: University of Virginia, 1997). See also his webpage at: http://www.hawaii.edu/powerkills/.
55. Mark Levene has repeatedly pointed to crises as a factor in the development of genocidal situations. See for example his multi-volume *Genocide in the Age of the Nation State* (London: I.B. Tauris, 2005).
56. See for example Harad Welzer, *Selbst Denken. Eine Anleitung zum Widerstand* (Frankfurt/M: S. Fischer, 2013).

Subsistence societies, globalisation, climate change and genocide: discourses of vulnerability and resilience

Mark Levene[a] and Daniele Conversi[b,c]

[a]*Faculty of Humanities, University of Southampton, UK;* [b]*University of the Basque Country UPV/EHU, Leioa, Spain;* [c]*IKERBASQUE, Basque Foundation for Science, Bilbao, Spain*

Anthropogenic climate change poses the possibility of total human extinction. Subsistence societies, however, have been threatened with extinction primarily as a consequence of systemic development for a very long time. Recent genocide scholarship, more particularly in relation to indigenous peoples, has engaged with some of these issues, even while terminologies such as ethnocide, cultural genocide, and indigenocide may suggest a restricted field of vision. Here, we argue that the very nature of a neoliberal globalisation and concomitant nation-state building makes *all* subsistence societies vulnerable to what amounts to structural genocide. But how does climate change exacerbate or complicate this bleak picture? The political economy of 'business as usual' in its dialectical relationship with the biosphere (expressed in the rising concentrations of greenhouse gas emissions) poses an acceleration of subsistence society vulnerability with catastrophic potential for extreme violence. But another scenario also presents itself. The very ongoing, seemingly impossible existence of non-marketised societies in direct relationship with nature, poses the possibility of *their* resilience in the face of climate change rather than those operating according to standard globalised norms. In conclusion, we propose that the crisis of anthropogenic climate change directly challenges not only assumptions about the 'inevitable' trajectory of globalisation with its supposed cast of survivors and victims but more precisely the purposefulness of 'techno-rational' epistemologies as set against those which might help humanity recover the possibility of a 'moral economy'.

Introduction

Trying to get one's head round the aetiology and morphology of anthropogenic climate change is a little like doing the same for genocide, except *much* more so. We might begin with the recently coined notion of the Anthropocene. The climate change equivalent to Raphael Lemkin's big idea – in this case, with the responsible neologists, earth scientists, Paul Crutzen and Eugene F. Stoermer[1] – accepting the proposition would be to acknowledge that humanity writ-large now constitutes a geological actor on the planetary stage with the ability to change 'the most basic physical processes of the earth',[2] albeit involving a trajectory towards epochal terminus so rapid that the Anthropocene will register as little more than a blip in earth history. Humanity, at least as we have normatively come to understand it since the onset some 10–11,000 years ago of the last great, *natural* shift at the Pleistocene-Holocene boundary, has always operated within 'parametric conditions', in other words a *relatively* stable climate which (mostly) year on year, millennium upon millennium, has allowed for cereal grasses to grow and for humankind to prevail in large numbers.[3] By contrast, the Anthropocene, whose beginnings c. 1750–1800, can be quite precisely dated in terms of increasing (previously constant) carbon emissions into the atmosphere as a direct consequence of the anthropogenic burning of fossil fuels (i.e. sequestrated solar energy from earlier geological eras), is now having biospheric consequences so profound that before the planet re-balances its own dynamic thermal equilibrium a great extinction event, as has happened at other geological boundaries, cannot be ruled out.[4] This would suggest an appropriate terminology not of genocide but *omnicide*. Or to put the matter tersely, in the light of what the science is telling us, the entirety of the 'developed' world's standard operating procedure encompassing economics, technologies, socio-cultural behaviour, not to say fundamental value systems, can no longer be sustained as viable or beneficial for ourselves, let alone for the planet's many millions of other species upon whom we also fundamentally depend. The knock-on effects for all academic disciplines, genocide studies included, are equally stark. Either we rethink *some* of our first principles, not to say standard modus operandi in the light of anthropogenic climate change, or we literally consign ourselves to the dustbin of history.

If this is a case of throwing down a gauntlet to fellow colleagues, it also carries with it an acknowledgement that the genocide field is somewhat unusual in the degree to which detached clinical observation often carries over into a desire to 'do something' most obviously attuned to practical prevention against any number of crimes against humanity. That particular boundary may be elastic, even fuzzy. However, part of the challenge here, is not simply to review some of the basic theoretical assumptions we often take as 'givens' for genocide studies, but to translate those reformulated ideas into a broader stream of critical thinking which – in repudiating 'business as usual' – seeks to find practical, non-violent paths towards human survival and well-being for *all* people, across the breadth of the planet, and for the long-term.

In this discussion piece, therefore, one of the things we are seeking to do is run with some familiar 'genocide' terminology, but to radically recast it in order to delineate some of the power relationships which are making some peoples on the planet ostensibly more vulnerable to climate change than others, and, by implication, more threatened with violent extinction. Holocaust research has given us the notion of 'perpetrators' and 'victims' (as well as 'bystanders'). The very terms may be debatable. But, just for argument's sake, we use them here to develop a distinction between those most and least responsible for climate change. It is certainly a very unequal relationship. While the emerging middle classes in all rapidly developing countries, China and India in the lead, would fall into the first category, the lion's share of carbon emissions can be directly or indirectly attributed to the lifestyles, working and consumer patterns of people in the First World,[5] a category, incidentally which would include nearly all, if not all, genocide scholars.

One might forcefully object that describing any population segment as perpetrators by dint of their carbon footprint is not appropriate, proportionate or kind. Perpetrators are people who kill and harm others, usually (though not always) knowingly. By contrast, the energy use of ordinary First World people is not so much consciously determined but rather predetermined by the function of the society they live in. If this leads indirectly to rising sea levels and methane emissions, glacial or Arctic melt, increased drought or flooding, deforestation, the death of the oceans, and so on, the evidence for *intentionalist* malice may still be absent. On the other hand, just as in cases of genocide where individuals and communities 'go along' with a regime actions because they are unwilling to question its behaviour, or possibly fear the consequences, so here on the climate issue a general Western failure to change addictive behaviour associated with our normal quotidian consumerist activity does raise questions about the degree of 'ordinary' complicity with what amounts to systemic violence against people and planet.[6]

If the notion of 'ourselves' as climate perpetrators will undoubtedly excite comment, far fewer are likely to contest that the Third World's very poor are its most obvious 'victims'. What particularly interests us here, however, are those subsistence societies at this end of the spectrum whose modus operandi involves almost no carbon emissions whatsoever. Poor and subsistent are not necessarily one and the same thing. Indeed, while billions of poor may be dependants of the global system, what characterises subsistence *cultures* is the manner in which they have traditionally created a high degree of socio-economic autonomy. This has been largely achieved through deploying their own labour or that of domesticated animals, aligned with a usually conservation-minded utilisation of whatever renewable energy sources are available to them for their basic sustenance.

Readers should see that what we are setting up here is a double-edged dialectic. The threat to the biosphere is symptomatic of a particular industrialising trajectory in human development, which we usually speak of as 'progress', and which, in the short-term, has favoured 'our' Western-led planetary predation. Putting to one side the important debate as to whether a wider range of subsistence cultures has modified the climate over the *longue durée*, in Anthropocene terms, their impact, by comparison with our own, has been negligible.[7] But viewed through the prism of self-sufficiency and sustainability, the argument is arguably reversed. While subsistence societies are not all alike but actually very different – itself indicative of long-term local adaptations to specific ecological niches – what they have in common is their direct *productive* relationship to habitus and hence nature. *Homo anthropocenus*, by contrast, is entirely dependent as a consumer within a globalised, inherently non-sustainable monoculture – one which takes from the global commons but gives nothing back in return and when it finally breaks down, as it *must* under the weight of this negative equity, will have no obvious access to, or ability to work *with*, nature for survival, let alone renewal. It is this tension between profligate consumers and productive subsisters as either vulnerable or resilient that is at the heart of our discussion.

Subsistence societies: the vulnerability of history, the evaluation of genocide scholars

To lump together peasants, nomads, fishing and swidden communities, foragers, pastoralists and hunter-gatherers, ought itself to excite comment. Granted that many of these communities have traditionally practised a mix of subsistence techniques, any immediate similarity between, for instance, Inuit and Masai, Tuareg and Yanomani, the Ijaw fishermen of the Niger delta, or the Banaue terrace rice cultivators of the northern Philippines should automatically be negated by simple contrast between the range of lived practice and cultural behaviour as operating within entirely distinct ecosystems. Lévi-Strauss-style structuralism notwithstanding,[8] the ability to survive – and cross-generationally change – is a function not of subsistence societies' sameness, but marked ecologically adapted versatility. So, why consider them as a whole? Two important reasons present themselves in the context of this discussion. The first we have already hinted at: their commonality lies in the fact that none of them have traditionally survived except in direct, physical relationship to the land, seas, forests or waterways and their respective *biota*. The primary concern of all such communities thus is to extract enough food and fuel to live from season to season, year to year, generation to generation, any surplus as such being exactly that: something in the first instance to be conserved or stored for future moments of scarcity or crisis. This does not mean that subsistence societies either live in isolation from one another or from thoroughly marketised societies. Across time and space, barter and trade has been a normal function of *all* societies. The subsistence variety does not live entirely outside or apart from the market. The point is one of priorities; the notion of economy for subsistence societies means first and foremost providing for the needs of the household from and through nature.[9]

But it is exactly such priorities which the modern marketplace rules out of court. The now-accelerated drive towards the commodification of everything within a global, capitalist, political economy, has no truck or time for ecological versatility except in so far as it can be quantified in terms of monetary value. As for what the biosphere itself provides for human need, for standard economics these are no more than externalities with no tangible 'value' whatsoever for contemporary, economic forecasting.[10] In a critical sense this tells us why not only ecocide is becoming inexorable, but is also inextricably linked to the destruction of subsistence cultures.[11]

We picked the names of six such diverse groups above quite at random: yet each is faced with the imminence of foreclosure either directly, because market forces have been grounding down their ability to survive outside the system, or indirectly, because these same forces have damaged, polluted, despoiled or sequestrated their *habitus* to the point where its natural resource base is no longer *replenishable.* In each instance, climate change is becoming a demonstrable and increasing factor in this negative equation. However, first cause is a political economic system that, as it has arisen to universal hegemony, has claimed for itself a sole monopoly over the global commons. As a result, not only is the entirety of nature disposable to *its* needs, but so too are alternative, local, human ecosystems whose reluctance to forego *their* existences or, worse, refuse to submit to globalisation, makes them obstacles in the path of progress and, hence, ripe for 'legitimate' erasure.

This perpetrator rationale will not be lost on genocide scholars, especially those who have charted Western colonial expansion. It is noteworthy that some of the most acute analysis has been of the 'Anglo' conquest of Australia. Here the ensuing settler encounter with the continent's aborigines, by the very nature of its usurpation of native lands and livelihoods regardless of any conscious intent *inevitably* led to what has been described in Marxist terms as 'relations of destruction', or indeed, as a specific 'process' of *indigenocide.*[12] Such assessments particularly resonate with our argument, not only because they offer a historical sub-set of a wider natural historical disaster viz. the 'ecological imperialist' assault on the fauna and flora of Australia,[13] but also because they can be cross-referenced with a broader metahistorical consideration of the Anthropocene. Here, after all, was arguably the starkest of cultural collisions at the onset of the modern world: one part of which consisted of peoples who had lived in largely unchanging ecological balance with their environment for not less than 50,000 years, the other part of which constituted an extrusion of peoples from half way across the planet at the very moment when *their* industrial revolution was wrenching the biosphere into an altered state. It followed that Anglo settler aims had nothing to do with sharing an actually very fragile ecosystem for subsistence purposes, but engrossing it as rapidly and rapaciously as possible, in order to create and then maximise its agro-business potential for the global market. One rapid anthropocenic consequence was the huge jump in Australia's methane contribution to global warming, thanks to the wholesale turning over of conquered territories to livestock production.[14]

Viewed through this Anthropocene prism, however, this destruction process, while still dialectical, has a further exquisite twist. From an earlier generation of genocide scholars, a farsighted Richard Rubenstein argued that the relationship between genocide and modernity can be traced back to the new economic relationships arising out of the English enclosure movement and the displacement of a hence superfluous peasantry overseas to new colonial habitats including Ireland and Australia.[15] Putting to one side the deeper ecological crisis of feudalism, out of which these new economic relationships emerged,[16] one might be inclined to dispute the Rubinstein thesis on the grounds that while an English peasantry was almost entirely dispossessed, often criminalised and repeatedly deported en masse as penal settlers to the colonies, it were *not* exterminated. But viewed anthropocenically we might equally propose that, not only is the distinction between Anglo and aboriginal subsistence cultures less sharp, but that their shared tragedy was to be victims of an emerging hegemonic, fossil fuel energised, Smithian-Ricardian 'economy of profit' which in the process turned English peasants into aboriginal exterminators.

Even more exemplary, one might argue, is the case with the former's northern counterparts in the Scottish Gaeltacht. This truly ethnographically distinct, autonomous, tribally-based and almost entirely subsistence society was 'cleared' to make way for Cheviot sheep, thus confirming that the destruction of native cultures in the interests of 'Anglo' agro-business could be enacted simultaneously and without notable racial differentiation on both sides of the globe.[17] Once again, however, one set of subsistence victims – evicted Scottish highlanders – played a notable role in the destruction of another indigenous set on the frontiers of Anglo expansion, Australia included.[18]

Is this the point at which to propose that there might be something structural in the nature of modern genocide, as a by-product of its very developmental trajectory? And that this in turn might act as a window into the *violent* origins of climate change? Lemkin, by dint of writing in an earlier time and with a focus primarily on nation-state versus 'minority' communities, may have only dimly conceived the connection. However, a recent piece by Adam Jones does pose the relationship between structural violence and genocide. Jones adumbrates a range of examples where 'core structural inequalities in the international system', or more precisely the distinction between the rich Western (or northern) world and the poor South, year on year, lead to the misery, debility and the premature or violent death of tens of millions.[19] He does not fully spell out that most of this death is itself an outcome of a developmentalist logic, underpinned by a contemporary neoliberalism whose ethos and practice effectively denies subsistence cultures their wherewithal outside, or apart from, the global market. Nor does he explore, except in passing, how the biospheric consequences of this accelerating trajectory are eating away at the parametric conditions essential to subsistence survival. That said, by way of Dirk Moses, we might be able to offer some linkage between Jones' structural violence proposition and Lemkin's cultural grounding to his genocide premise.

Moses has repeatedly insisted that Lemkin's reference to *cultural* genocide was not some dispensable afterthought but fundamental to his exposition of his entire legal-preventative framework.[20] Through this perspective we might read Lemkin's proposed instrument as an attempt to quell the political violence inherent in the modern, monocultural nation-state as directed against diverse but disempowered ethnic or ethno-religious pluralities. However, it would also be difficult to detach the notion of culture from most subsistence societies, not least given their relationships to native habitus and, hence, discrete ecological adaptations as sustained and developed over generations. In these terms, any culturally homogenising and/or economically driven project that might seek to violently disrupt or suffocate traditional, diverse, subsistence practice could also be taken to be genocidal.

The most obvious and devastating example would be Stalinist collectivisation, which, by dint of its overt intent to physically liquidate a top tier of Soviet peasantry, branded by the authorities as 'kulaks', offers a rather convenient confirmation of these events as genocide in a more standard sense. However, as a recent study illuminates, this was not a selective attack on only part of the Soviet peasantry, but rather a war against the peasantry *tout ensemble*, aimed at the complete destruction of an entire set of subsistence cultures.[21] The fact that this was ordained by a communist regime should also give us pause before ascribing to capitalism sole responsibility for such ruthlessness. Instead, the connecting thread between the Soviet industrialising rupture and the Russian *longue durée* had been anticipated by the somewhat longer, if still quite rapid, destruction of Britain's peasant base at plutocratic hands, culminating some 100–150 years earlier. Through an Anthropocene lens, these cataclysmic events might be understood as part of a series of environmentally Pyrrhic victories by elites who, blinded by their conviction of the possibilities of a rationally based terrestrial dominance, ascribed to themselves the role of progressive modernists against the antediluvian inertia of a peasant 'savagery'.

In this manner, *modernism*/modernity has appropriated to itself the notion that it is always on the right side of history. As an all-pervasive institutionalised ideology indeed, it can invoke 'the progress of the nation' to make any argument to the contrary 'luddite' and so inadmissible on grounds of its alleged, backward-looking refusal to bow to the 'inevitable' and *good*.[22] As a result, it is almost inconceivable to imagine what the world might have been like if Western subsistence societies had been allowed to survive, even flourish. Or, for that matter, where we would be today in terms of climate change.[23] Such counterfactual musings aside, it was the subsisters themselves who paid the consequences for the mandated current of history, whether they actively resisted it or not. A hundred years on from 'Captain Swing' and the last gasp of English peasant opposition to the new hegemony, a repeat sequence of epic droughts swept *fin-de-siècle* Asia, Africa and Latin America. Climate science now knows that they were part of the signature of a periodic shift in tropical weather patterns otherwise described as ENSO: El Niño-Southern Oscillation. At least 30 million subsistence peasants starved to death or died from epidemics as a result.[24] But it was not a climatic shift *alone* that did for them. It was that the doctrinaire laissez-faire diktats, economic wisdoms and taxation regimes of the new imperialism (formal or otherwise) ensured that traditional practices and technologies of conservation and replenishment, which *normally* would have come into play to feed the vulnerable, were prevented from doing so. Was this genocide? Not in the exact sense we have come to understand the term. But, structurally speaking, the late-Victorian famines were the 'forcing houses and accelerators of the very socio-economic forces'[25] which helped to destroy the stability and resilience of peasant societies world-wide. In a trice, these forces created what we now so egregiously denote as 'the Third World', while setting humanity on its relentless path towards, not just massive global inequality, but *sustained* biospheric emergency.

The situation now: how does the high Anthropocene heighten subsistence societies' vulnerability?

One might argue *plus ça change, plus c'est la meme change*. Just as usually affluent people are inclined to intone biblical verse to the effect that 'the poor will always be with us', by the same token one might propose that peasants, nomads, and the like, have always been vulnerable to the power of states, seigneurs and economic asset-strippers, so long as there has been civilisation. Update the *dramatis personae* to include transnational corporations, government technocrats and International Monetary Fund planners and it remains the same old story of the weak making way for the strong, preferably accepting their misfortune or even extinction without a fuss, while the rest of 'us' look on, perhaps comforting ourselves as did white Australians not so long ago, when they hoped that something might be done for the aborigines 'to smooth the dying pillow'.[26]

Except now there is qualitative and quantitative difference. To be sure, we would not need climate change *per se* to confirm the acute threat to the existence of *all* subsistence societies. An incipient literature on the Friedmanite-informed deregulation of the 1970s and 1980s makes abundantly clear the relationship between the rise of neoliberal globalism and Third World debt, displacement and degradation, environmental or otherwise.[27] Against this merciless juggernaut, international efforts to protect what residually remains in the hands of native cultures, such as the 2007 UN Declaration on the Rights of Indigenous Peoples, not only lack teeth but are unlikely of themselves to stop predatory states or corporates in their tracks from wresting whatever fossil fuels, minerals, timber or water resources may lie on or beneath native soil.[28] It is this very process of globalising overdrive which is now pushing the biosphere over the brink, but with the frontline in the assault the very same subsistence cultures whose dependence on some basic predictability, not to say sustainability of our 'parametric conditions', remains absolute. The problem is that such conditions no longer pertain.

The Intergovernmental Panel on Climate Change (IPCC) – the United Nations (UN)-supported, consensual and, thus, in its published findings actually rather conservative, scientific body which assesses and summarises the current state of climate change research – has identified what is climatically known as the Intertropical Convergence Zone (ITCZ). This is the arena in which the most deleterious impacts on human populations are already taking place and predicted to rapidly worsen as greenhouse gas emissions continue to rise.[29] The ITCZ, as its name implies, embraces the tropics and with them the vast majority of the remaining subsistence societies. And putting aside underlying indicators of climate change, such as increasing surface temperature and sea-level rise, the most immediate bad news for these subsisters is the seasonal shortening and weakening of annual rains; for instance, in the Indian sub-continent monsoon cycle. Paradoxically, this is often accompanied by a dramatic *increase* in annual precipitation in which truly torrential deluges can lead to the sort of catastrophe visited on Pakistan in July 2010 or the Philippines in November 2013. There should be no need to reprise here the long-term impact of these floods on a country like Pakistan, where serious malnutrition among some of the original 20 million displaced people – mostly peasant farmers – is still rife.[30] But ironically, while Western observers might be ephemerally aware of these extreme weather events through headline media, they are much less likely to be cognisant of the ongoing and cumulative catastrophe of drought, which, as in neighbouring Afghanistan, has been the 'invisible' crisis of the last *40 years*, and to which all other issues of insurgency and counterinsurgency, while closely related, are essentially ancillary.[31] We might at this point add in a host of other related climate factors such as the effects of deforestation and glacial melt, but the human matter in immediate question is how sustained climate emergency determines that 'subsisters' eat what remaining seed they otherwise would put aside for next year's harvest or, among pastoralists, slaughter the cattle which provides them with just about everything, or yet again, among river delta and shoreline communities who find that there are no longer any fish to be fished.

How can genocide scholars get to grips with what this long-term absence of food means for family groups and *communities*, let alone all the other psychological, emotional and cultural ramifications? Should we be reminding ourselves of the calorific intake of Warsaw ghetto inhabitants in the years 1940–1942? Or, perhaps, of how some peasants descended into cannibalism during the Ukrainian *Holodomor*?[32] The latter case is arguably more apposite in the sense that it gives some indication of what subsisters in a famine situation do: sitting tight in the first instance, eating domestic animals, next year's seed, weeds, indeed, whatever they can, and only in utter extremis abandoning hearth and home. But it is interesting to consider how just as rural access to towns was denied with lethal effect by the Soviet authorities, the primary *leitmotif* of US and other advanced states' response to climate change has been to plan for how most effectively to keep the 'barbarians at the gate'.[33] In fact, as one recent assessment makes clear, it is unlikely even under the most severe pressures that the majority of famished subsisters will seek, or have the wherewithal to move as far as, the heavily barricaded borders of the planetary plunderers. Much more likely is that movement, where it occurs, will be in the famished subsistence zones themselves, leading to intense inter-group conflict for remaining water holes, cattle, and whatever land which is not so desiccated as to produce some meagre crop.[34] Some of this type of interaction has impinged on genocide scholarship by way of Darfur, though more seasoned East Africa observers will know that acute climate variability has always been a factor setting different groups of pastoralists and farmers against one another. Recent studies, however, suggest that the situation in particularly challenged regions is getting much worse; and not just in East Africa. Large swathes of central Asia as well as Central America are unravelling into conditions of extreme, endemic violence.[35]

It might be convenient to blame this descent on specific climate change impacts in these regions or, more conveniently still, on the supposed propensity of particular societies to Kalashnikov-assisted, warlord rule for which the blame can then be levelled at 'failed' states. The value of the latter explanation lies in the way it distances First World actions from responsibility for *anything*. But monocausality equally misses the point about climate change that *nearly* always 'unfolds as part of a matrix of causality'.[36] Pentagon speak talks of climate change as a 'threat-multiplier'. While accepting the general point, genocide scholars ought to be wary of where this sort of analysis is coming from: US and allied security 'think-tanks' being much keener to address the 'threat' of Third World collapse to 'us', rather than attempting to give serious consideration as to how an orderly retreat from the fossil fuel-driven global economy might be accomplished.[37]

In fact there *are* instances where climate change by itself represents a primary and direct threat to some indigenous societies, particularly those whose livelihoods are dependent on the *icebound* migration of semi-wild reindeer (caribou). The place where global warming is advancing at its most alarming rate is not actually in the ITCZ but in the Arctic north. And while we do not have space to develop a chapter and verse explanation of how Arctic melt is having disastrous consequences on the well-being not just of large mammals, but of an entire ecosystem, it should be obvious why, as reindeer numbers collapse, the impact on otherwise extraordinarily well-adapted herding communities, such as the Saami of northern Scandinavia or the 5000 Mongolian-related Dolgans still living on the remote Taimir peninsula in the far northern reaches of central Siberia, is looking decidedly bleak.[38]

If the fate of such peoples should focus our attention on the fragility of *all* humankind in the face of biospheric blowback, the psychological and actual remoteness of small, sub-Arctic cultures may provide an obstacle to 'our' full appreciation of their particular dilemma.[39] Arguably less marginal to Western cognitive maps but also closer to how climate change is accelerating toxic, developmental factors already strongly disintegrating a range of subsistence societies are provided by the leading Indian environmental activist Vandana Shiva. It is Shiva who, without apology, uses the term 'disposable' to describe how the modernising 'monocultural' mind set treats subsistence cultures.[40] Shiva also makes no bones about the *value* of subsistence societies for combating climate change and feeding the world population. Yet on the sub-continent, peasants are under relentless attack. The first cause appears to be the increasing sequence of long-term drought, punctuated by catastrophic flooding. Shiva writes of the Bundelkhand region between Uttar Pradesh and Madhya Pradesh where, as 'agriculture has virtually come to a halt', not only starvation but also suicide is endemic. In fact, this situation has become rife across much of central India where official estimates of 200,000 rural suicides over the last decade, on average one every 30 minutes, may be an extreme underestimate given that family members are often not included in the fatality figures.[41]

However, as Shiva and others have demonstrated, drought is, inaptly speaking, only the tip of the iceberg. The real crisis began in 1997, when having previously encouraged peasants to switch to cultivating a cotton cash-crop, the Indian government then dropped the subsidy. At the same time, the US transnational Monsanto introduced genetically modified seed heavily backed as drought-resistant but, in practice, very dependent on artificial fertiliser and pesticides for success. Equally telling, from a peasant perspective, is the fact that because the seed is a hybrid it cannot be saved from the previous crop for re-use but *has to be* re-purchased. In short succession, subsistence farmers found themselves producing something which they could not eat, demanded crippling quantities of scarce water to grow, rapidly degraded generation upon generation of carefully nurtured soil fertility and then promptly failed when drought kicked in. However, the first cause of the farmers' woes was *not* climate change, it was the structural violence visited on them by Delhi's embrace of globalisation, combined with Monsanto's suave purveyance of what amounted to agro-industrial dependency.[42] Having given up their social and cultural capital to these forces and been forced foursquare into full market economy obligations over which they had no control, the further stations of the subsisters' cross were enacted when massive debt incurred in order to make the switchover in the first place could not be repaid to unscrupulous *zamindaris* and moneylenders, and against which the state was meant to protect them. This ensured that a return to sustainable micro-husbandry also became impossible. Only three options at this point thus remained: eviction, self-induced flight (to the already burgeoning mega-slum ghettos of India's cities – some *20 million* hungry peasants decided on taking this route in the last decade) – or suicide.[43]

A case of genocide? Clearly not, at least not according to the conventional wisdom. A further example of how we are perpetrators and they are victims, indubitably. Because this is not just about the destruction of peasant autonomy, the loss of biodiversity, or even with it, the massive hole in carbon sequestration which this high entropy agro-industrial disorder has brought in its wake. What is entirely less visible is the massive carbon footprint from taking raw cotton to be manufactured, usually in exploitative sweatshops (the labour again very often being displaced subsisters) and then transported again, all over the world, so that we, already affluent consumers, may purchase *cheap* cotton clothes before finally squandering a whole lot more CO^2 machine-washing them.[44] The last point may sound throwaway. Instead, it should reinforce that manner in which our political economy is ordered to convenience a largely First World section of the global family at the, albeit several steps removed, exterminatory expense of Third World subsisters.[45]

A further brief example is relevant here by way of the scandal of biofuels. On paper, the increasing worldwide, government-backed conversion from oil to renewable biomass – as part of the route by which current energy demands can be ensured while reducing carbon emissions – would seem laudatory. The practice is quite different; the rich West, including the European Union, is effectively subsidising Third World countries to either destroy millions of hectares of virgin rain forest to plant biofuel crops or to appropriate subsisters' habitus for the same purpose. Leave aside the studies demonstrating that the net result is *more* greenhouse gas emissions, strain on water supply, soil depletion, not to say spiralling food prices, as Third World land is switched from food to fuel production.[46] Instead, let us concentrate on the immediate impact on subsisters. Again, our source is Shiva and her focus is the tribal peoples (*adivasi*) of Chhattisgarh and Rajasthan, where traditional food security founded on rice production has been supplemented by customary grazing rights on the village commons. It is these very commons, however, that the Indian government sees as a major arena in its drive to develop the biofuel crop, *jatropha*. The government's planning commission describes the land in question as 'unused' or 'waste'. We are literally back where we began with the forcible eviction or marginalisation of the English subsister, the difference being that the current high Anthropocene phase in the enclosure of the global commons is now favouring not just rich, urban, Western people but a whole tier of globalised consumers who either cannot see or do not understand how their lifestyles are driving not only climate change but the most vicious, structural violence against subsisters everywhere. As Shiva succinctly puts it: 'cars will run while animals and people starve'.[47]

One might rejoinder: how can global consumers know what is going on? It is a function of our digital-age media that even the ongoing, sometimes very violent aspect of subsister resistance, and the massacres which ensue, are either strikingly underreported, or, as in Chhattisgarh, easily blamed on left-wing Naxalite extremism.[48] But, should we really expect otherwise, given that it is not just India's internal colonialism which is at issue here but, behind it, some of the most powerful political and economic forces on the planet striving to keep afloat a system which, biospherically speaking, is entirely redundant?

Subsistence societies as champions of resilience

One thing which is notable in hegemonic discourses about subsisters is how little the language has changed in 200 years. What was 'improvement' and 'progress' then is simply 'sustainable (sic) development' now, while those who insist on doing things the old way are clearly mired in their own uneducated obduracy and backwardness. Three or four decades ago, when there was a vogue for peasant studies, René Dumont recounted with almost perverse pleasure the story of the old Auvergnais peasant woman, who 'as late as 1936′ was gathering wild hazelnuts from hedgerows, shelling them and after two months of spare-time work producing three or four litres of wild oil, 'her only concession to progress the use of an old mechanical press'.[49] Dumont seemed to concede the woman's 'mere collecting' might be quaint but otherwise entirely antiquated. But through knowing what we know today, might we not offer a very different assessment? After all, here was someone working through an indigenous knowledge of her local woodland habitat, enabling her year on year to harvest a crop without damage to the environment, and then by the sweat of her own brow, producing a highly valuable product without any carbon emissions whatsoever. To be sure, there is a temptation in reading into such practice some romanticised notion of a prelapsarian paradise. Subsistence existence was never and has never been anything of the sort. Over and beyond the sheer, monotonous hard grind of daily backbreaking toil, peasant life in particular has always had elements of hide-bound small-mindedness, and certainly its own very visceral quality.[50] Nor has living close to nature always meant getting it right; cases of ecological maladjustment or worse, rampant despoilation, littering both the deep historical and more recent historical landscape.[51]

However, *much* of this litany of disaster can be laid at the door of *civilisations*; in other words, more complex societies, rather than those operating at grass-roots subsistence level.[52] Part of the enduring quality of the latter is bound up with, on the one hand, constant anxieties about the unpredictability of nature and the potential scarcity that carries with it, and, on the other hand, a recognition that avoidance of dearth means both working with nature (as indeed with one's neighbours), to ensure its sufficient regenerative capacity for the next season, the next generation. This is the reason why we can speak, with Fernand Braudel, of the *long durée*.[53] And why the discontinuity with our own dominant political economy is so jarring.

Shiva has described this disjuncture as between a self-organising, self-renewing autopoetic system, founded on self-generative endosomatic energy and that of a mechanistic, allopoetic system founded on external, exosomatic sources. The contrast might equally be described as that between a flexible heterarchy and brittle hierarchy, the former providing for both ecological and human diversity, the latter for monocultural conformity. It is significant, notes Shiva, that the allopoetic labels its opposite either 'undeveloped' or 'underdeveloped'.[54] And to be sure, because subsistence cultures as fundamentally autopoetic work within and are dependent upon the seasonal and cyclical rhythms of nature their mind set is one of deferred gratification. The good life, for them, necessarily means, for instance, not consuming the 'instant' yet perfect globally sourced meal, but rather *conserving* diverse seed varieties derived from local soils and microclimates. The very nature of such long-term cultivation requires meticulous nurturing and goes hand in hand with what Shiva calls 'grandmother knowledge'. Translated into a broader vision of a resilient future, Shiva's earth democracy is particularly, if not defiantly, woman-focused. And Shiva's own Navdanya biodiversity and organic-farming movement has shown through repeated practice that fossil fuel-free, GM-free, sustainable, agro-ecological techniques, using for example saline-resistant seed varieties to counter the effect of climate change, can lower the costs of production, produce *more* food yet without 'destroying the environment and killing peasants'.[55] But it cannot be replicated or repackaged for the instant 'green revolution' (sic) fix. The contrast, thus, at it starkest, is between a localised, women-orientated continuity and big solution, intrinsically 'male', globalised rupture, the end game of which can only be mass, anthropocenic violence.

A further example of this collision between life-giving and death-dealing cultures is presented in the British environmental activist Nicola Peel's documentary film, *Blood of the Amazon*.[56] The blood in question actually refers to the 18.5 billion tons of toxic waste spilled into the Panacocha region of Ecuador's Yasuni national park, over the last 30–40 years, as the 'price' for Chevron-Texaco's major oil and gas production programme there. For the record, this amounts to a spillage 30 times greater than that of the Exxon-Valdez disaster, which disastrous enough in terms of Alaska's wildlife, is belittled by the impact on the supposedly 'pristine' Yasuni rain forest, a mere 0.2% of the planet's *terra firma*, yet with 25,000 different plant species alone, its most bio-diverse habitat. Nor is it just the flora and fauna that have suffered. Put aside the million metric tons of carbon dioxide emitted into the atmosphere, at ground zero the Panacocha's contaminated soil and water translates not only into daily illness but 'an epidemic of cancer, miscarriages, birth defects' and the like for its five groups of indigenous people. Indeed, the mortality rate currently runs at 143 deaths for every 1000 births.[57] Thus far, the Yasuni story sounds like just another addition, if a particularly egregious one, to the litany of 'sacrifice zones' developed for the unconscionable benefit of corporate interest; its *indigènes*, the hapless victims as well as bystanders to their own fate.[58]

Yet the other side of the story is rather different. It speaks of people resistance *and* resilience. The very fact that from being originally cowed by both Chevron and their frontline Ecuadorian military accomplices, the Cofan, Siona, Secoya, Kichwa and Huaorani have come together as a federation to mobilise against further encroachment, symbolised in the non-violent defence of Sarayacu, in 2002, when the community's women prevented soldiers and oil workers from invading their land.[59] Better known internationally is their success in the Ecuadorian courts where they have forced Chevron to stand trial. In a classic court ruling in February 2011, the company was found liable for $8.6 billion for environmental clean-up costs.[60] This dovetails with the aspirations of the present left-leaning government of Rafael Correa to become a post-petroleum economy. While this promise has yet to be fulfilled, the communities themselves have embarked on a decidedly post-petroleum 'development'. On the one hand, they have brought in experts in *mycoremediation*, a biological technique to help sponge up the soil and groundwater contamination. On the other hand, in the area along the Napo river, they have sought not only to reinvigorate but to share with outsiders their native craft, fishing, permaculture and herbal medicinal skills, intrinsic to living *and flourishing* in and from the forest.[61]

How much can such subsistence cultures teach *us*? And is this just a matter of the transmission of practical Amazonian or Arctic skills, which, otherwise, may not seem terribly relevant? Let us get down to some fundamentals. In 2006, the British Met Office's climate research wing, the Hadley Centre, having collated copious data and observations, produced a climate model which forecast that fully one-third of the planet's land mass will be desert by 2100, while up to half of land surface will suffer drought.[62] The conventional wisdom nearly always assumes that the poorest, subsistence societies under these sort of pressures will be the first to collapse, with unmitigated violence the corollary. However, there is equally a case for arguing that the most vulnerable societies to this sort of scenario are the *most* advanced, developed ones such as 'ours'. Why? First, because of our almost total dependency on thin, often distant supply lines to provide basic services, including water, food, heat and light. Overwhelmed by some extreme, sustained weather event and the truism about hierarchic, complex, urban societies being only three or four meals away from anarchy could well be confirmed to disastrous effect.[63] But if this speaks to the physical realities before us, it critically misses the emotional, and what one might call psychic, element of the equation. With only a linear script of secularised, terrestrial mastery to hand, there is, paradoxically, an almost palpable absence, not just of any awe of nature, but alertness to Gaia's refusal to take any more violation. One result is a form of schizophrenia. On the one hand, there is assumption that whatever happens, there will be a technical fix to sort it. On the other hand, there is a set of semiotically informed stories we tell each other, particularly through fantasy and science fiction films, in which there *is* catastrophe, it is overwhelmingly apocalyptic, but to which the only response can be a highly individualised, social Darwinian struggle for survival.[64]

Compare and contrast with societies close to nature. Traditionally built into their very fabric is not just the wonderment but also an entirely 'normal' precognitive alertness of nature's unpredictable and capricious power. Subsistence societies thus carry within themselves the imminence of disaster at all times. It is there in their cosmologies – often only partially dampened by any monotheistic superstructure – ensuring that a *religious* interpretation of the world is ever-present. Subsisters know that to ignore or defy nature is to bring down its wrath on the whole community with cataclysmic force.[65] What to us may be superstitious nonsense, to them is the forearming of the *collective* whole so that, faced with the approaching storm, they are not just physically but mentally equipped to ride it out. To which, one has to ask, in destroying these cultures, are *we* simply cutting the umbilical cord which might yet be humanity's lifeline in the face of climate catastrophe?[66]

Violent nemesis or resilient future?

Genocide is a by-product of the modern international, political economy.[67] But underneath that is something else: an ongoing structural violence whose origins lie in the very rupture with traditional subsistence lifestyles as sustained over thousands of years. Patrick Wolfe has notably argued that the settler colonial assault on the Aborigines is a case of structural genocide.[68] Whether the wider, developmental, state-cum-corporate degradation, displacement and destruction of *all* subsistence societies falls in this same category may be discussable. However, viewed through the contemporary biospheric crisis, the inference takes on an altogether broader meaning. As Shiva, as always overturning standard wisdoms, puts it:

> In the biodiversity paradigm, to be developed is to be able to leave ecological space for other species, for all people and future generations of humans. To be underdeveloped is to usurp the ecological space of other species and communities, to pollute the atmosphere, and to threaten the planet.[69]

Shiva has no doubt that, putting all other things aside, the industrialisation of food 'has put the human species on a slippery slope of self-destruction and self-annihilation'.[70] Parenti has equally noted that rather than changing course in response to climate change, 'business as usual' is now operating on repression, surveillance and violence as its form of adaptation.[71] Not only can we expect this Schmittian-style state of emergency to become the 'new normal' as possibilities of climate mitigation recede, but with the richest, most powerful states moving to a further default position by way of geo-engineering.[72] All this implies not only a more pronounced role for military-industrial complexes as the defenders of the status quo but an accelerating, militarised scramble for what remains of the global commons.[73] Traditional subsistence societies necessarily will be the ones who suffer most in the path of this trajectory.

That said, we must reject any notion that this entails the vulnerability of these societies but not of ourselves. On the contrary, the globalising project that has brought us both genocide *and* climate change urgently demands interlocutors who can articulate an entirely different epistemology of human relationships to each other, as to nature. In a recent volume on this theme, Lewis Williams and his co-authors significantly argued not only for a repudiation of an overtly techno-rational approach to ecological dilemmas but a decolonisation of consciousness to accompany it.[74] We propose that scholars dealing with human rights, genocide and related areas need to be part of that 'decolonising' tendency: beating the path not only to solidarity with subsisters everywhere, but giving backing to the insight that climate resilience is founded on both our own recovery of native, practical skills and on that pre-Anthropocene virtue called the 'moral economy'.

Notes

1. Paul J. Crutzen, and Eugene F. Stoermer, 'The "Anthropocene"', *Global Change Newsletter*, 41 (2000): 1718. See also Will Steffen, Jacques Grinevald, Paul Crutzen and John McNeill, 'The Anthropocene; Conceptual and Historical Perspectives', *Philosophical Transaction of the Royal Society A* 369, no. 1938 (2011): 843–5, for antecedents to and emergence of the concept.
2. Naomi Oreskes, 'The Scientific Consensus on Climate Change: How Do We Know We're Not Wrong?', in *Climate Change, What it Means for Us, Our Children, Our Grandchildren,* ed. Joseph F.C. Dimento and Pamela Doughman (Cambridge, MA: MIT Press, 2007), 93. See also Mark Williams, Jan Zalasiewicz, Alan Haywood and Mike Ellis, 'The Anthropocene: A New Epoch of Geological Time?', *Philosophical Transaction of the Royal Society A* 369, no. 1938 (2011): 835–41, for a broader discussion of the Anthropocene debate.
3. Dipesh Chakrabarty, 'The Climate of History: Four Theses', *Critical Inquiry* 35 (Winter 2009): 217–18.
4. Anthony D. Barnosky et al., 'Approaching a State Shift in Earth's Biosphere', *Nature* 486 (6 June 2012): 52–8, for one of the most recent assessments to this effect, by a team of 22 leading earth scientists; and Johan Rockström et al., 'Planetary Boundaries: Exploring Safe Operating Space for Humanity', *Ecology and Society* 14, no. 2 (2009): 32, for the wider set of earth-system planetary boundaries which, once transgressed by anthropogenic pressures, are likely to lead to irreversible changes gravely deleterious to human existence.
5. See Aubrey Meyer, 'The Case for Contraction and Convergence', in *Surviving Climate Change: The Struggle to Avert Global Catastrophe*, ed. David Cromwell and Mark Levene (London: Pluto Press, 2007), 32, Figure 1.
6. Clive Hamilton, *Requiem For a Species: Why We Resist the Truth about Climate Change* (London: Earthscan, 2010), for a searing indictment. See also the discussion of Western societal responsibility by way of Holocaust analogy in Harald Welzer, *Climate Wars: Why People Will Be Killed in the 21st Century* (Cambridge: Polity, 2012), 13–18. On the issue of intentionality versus unintenionality, see Rebecca Hofmann, 'Culturecide in Changing Micronesian Climates? About the Unintenionality of Climate Change', *The International Journal of Human Rights* 18, no. 3 (2014): 336–49.
7. William F. Ruddiman, *Plows, Plagues and Petroleum, How Humans Took Control of Climate* (Princeton and Oxford: Princeton University Press, 2005), Chapters 7–10. However, see Erle C. Ellis et al., 'Used Planet: A Global History', *Proceedings of the National Academy of Sciences/ PNAS* (29 April 2013). http://www.pnas.org/content/early/2013/04/25/1217241110. abstract (accessed 11 November 2013), for a recent article proposing that the most significant earth system changes were caused by human populations over 3000 years ago.
8. Edmund Leach, *Lévi-Strauss* (London: Fontana, 1970).
9. Karl Polanyi, 'Societies and Economic Systems', in *The Great Transformation, The Political and Economic Origins of Our Time* (Boston MA: Beacon Press, 2001[1944]), Chapter 4.
10. Ulrich Loening, 'The Attitude of Human Ecology', in *Radical Human Ecology, Intercultural and Indigenous Approaches*, eds Lewis Williams, Rose Roberts and Alastair Macintosh (Farnham and Burlington, VT: Ashgate, 2012), 22.
11. See Martin Crook and Damien Short, 'Marx, Lemkin and the Genocide–Ecocide Nexus', *The International Journal of Human Rights* 18, no. 3 (2014): 298–319.
12. Tony Barta, 'Relations of Genocide: Land and Lives in the Colonisation of Australia', in *Genocide and the Modern Age*, ed. Isidor Wallimann and Michael Dobkowski (Westport, CT: Greenwood Press, 1987), 237–51; Raymond Evans, '"Crime without a Name": Colonialism and the Case for "Indigenocide"', in *Empire, Colony, Genocide, Conquest, Occupation and Subaltern Resistance in World History*, ed. Dirk Moses (Oxford and New York: Berghahn Books, 2008), 133–47.
13. Alfred W. Crosby, *Ecological Imperialism: The Biological Expansion of Europe, 900–1900* (Cambridge: Cambridge University Press, 1986); A.J. Marshall, ed., *The Great Extermination: A Guide to Anglo-Australian Cupidity, Wickedness and Waste* (London: Heinemann, 1966).

14. David I. Stern and Robert K. Kaufmann, *Estimates of Global Anthropogenic Methane Emissions 1860–1993* (Australian National University, 1995). cdiac.ornl.gov/trends/meth/ch4.htm (accessed 11 November 2013).
15. Richard L. Rubenstein, *The Age of Triage: Fear and Hope in an Overcrowded World* (Boston, MA: Beacon Press, 1983), esp. 34–59, 120–7.
16. Jason W. Moore, 'Crisis: Ecological or World-Ecological?', in *Depletion Design: A Glossary of Network Ecologies*, ed. Caroline Wiedmannad and Soenke Zehle (Amsterdam: Institute of Network Cultures, 2012), 34.
17. See Eric Richards, *The Highland Clearances: People, Landlords and Rural Turmoil* (Edinburgh: Birlinn, 2000).
18. See Don Watson, *Caledonia Australis: Scottish Highlanders on the Frontier of Australia* (Sydney: Collins, 1984), 165–7, for the Warrigal Creek massacre, one particularly notorious example.
19. Adam Jones, 'Genocide and Structural Violence: Charting the Terrain', in *New Directions in Genocide Research*, ed. Adam Jones (New York and London: Routledge, 2012), 132–51.
20. See A. Dirk Moses, 'Raphael Lemkin, Culture and the Concept of Genocide', in *The Oxford Handbook of Genocide Studies*, ed. Donald Bloxham and A. Dirk Moses (Oxford: Oxford University Press, 2010), 19–41, esp. 22–30, 37; Jennifer Huseman and Damien Short, '"A Slow Industrial Genocide": Tar Sands and the Indigenous Peoples of Northern Alberta', *The International Journal of Human Rights* 16 no. 1 (2012): 221–2, for further exploration.
21. Lynne Viola et al., ed., *The War against the Peasantry, 1927–1930: The Tragedy of the Soviet Countryside*, trans. Steven Shabad (New Haven, CT and London: Yale University Press, 2005).
22. Daniele Conversi, 'Modernism and Nationalism', *Journal of Political Ideologies* 17, no. 1 (2012): 13–34.
23. Mark Levene, 'Historians for the Right to Work: We Demand a Continuing Supply of History', *History Workshop Journal* 67 (Spring 2009): 69–81. The essay is itself an *homenaje* to the great English social historian, E.P. Thompson, whose Marxism notwithstanding, looked back to the 'moral economy' of the eighteenth century as set against the emerging 'economy of profit'.
24. Mike Davis, *Late Victorian Holocausts, El Nino Famines and the Making of the Third World* (London; New York: Verso, 2001), 6–7.
25. Ibid., 15.
26. Quoted here in Anna Haebich, '"Clearing the Wheat Belt": Erasing the Indigenous Presence in the Southwest of Western Australia', in *Genocide and Settler Society, Frontier Violence and Stolen Indigenous Children in Australian History*, ed. A. Dirk Moses (Oxford and New York: Berghahn Books, 2004), 276.
27. Naomi Klein, *The Shock Doctrine: The Rise of Disaster Capitalism* (London: Penguin, 2008); David Harvey, *A Brief History of Neo-Liberalism* (Oxford: Oxford University Press, 2005); Manfred B. Steger, 'Ideologies of Globalization', *Journal of Political Ideologies* 10, no. 1(2005): 11–30; Daniele Conversi, 'The Limits of Cultural Globalisation?', *Journal of Critical Globalisation Studies* 1, no. 3 (2010): 36–59, for a small clutch of such studies.
28. Moses, 'Raphael Lemkin', 39–40; Survival International monthly email updates. http://www.survivalinternational.org/news/enews (accessed 11 November 2013), for the ongoing litany of struggle.
29. IPPC 5th Assessment Report (AR5), 2014. https://www.ipcc.ch/publications_and_data/publications_and_data_reports.shtml (accessed 11 May 2014).
30. 'BBC News – Pakistan Floods', 8 February 2012. http://www.bbc.co.uk/news/special_reports/pakistan_floods/ (accessed 11 November 2013).
31. Christian Parenti, 'Drugs, Drought and Jihad: Environmental History of the Afghan War', *Tropic of Chaos: Climate Change and the New Geography of Violence* (New York: Nation Books, 2011), Chapter 9.
32. Timothy Snyder, *Blood Lands: Europe between Hitler and Stalin* (London: The Bodley Head, 2010), 46–52.
33. See Dave Webb, 'Thinking the Worst, The Pentagon Report', in *Surviving Climate Change*, ed. David Cromwell and Mark Levene (London: Pluto Press, 2007), 59–81; and Steve Wright, 'Preparing for Mass Refugee Flows, the Corporate-Military Sector', in *Surviving Climate Change*, ed. David Cromwell and Mark Levene (London: Pluto Press, 2007), 82–101. Also Parenti, Tropic, Chapter 2, 'Military Soothsayers'; Gwynne Dyer, 'The Geopolitics of

Climate Change', in *Climate Wars: The Fight for Survival as the World Overheats* (Oxford: Oneworld Publications, 2010), Chapter 1.

34. Foresight report, *Migration and Global Environmental Change, Future Challenges and Opportunities* (London: The Government Office for Science, 2011). https://www.gov.uk/government/publications/migration-and-global-environmental-change-future-challenges-and-opportunities (accessed 11 November 2013).
35. See Jürgen Scheffran, Tobias Ide and Janpeter Schilling, 'Violent Climate or Climate of Violence? Concepts and Relations with Focus on Kenya and Sudan', *The International Journal of Human Rights* 18, no. 3 (2014): 369–390; Andreas Exenberger and Andreas Pondorfer, 'Genocidal Risk and Climate Change: Africa in the Twenty-First Century', *The International Journal of Human Rights* 18, no. 3 (2014): 350–68; Parenti, *Tropics*; Welzer, *Climate Wars*; and Dyer, *Climate Wars*, for the broader canvas.
36. Parenti, *Tropic*, 184.
37. Mark Levene, 'The Apocalyptic as Contemporary Dialectic: From Thanatos (Violence) to Eros (Transformation)', in *Future Ethics, Climate Change and Apocalyptic Imagination*, ed. Stefan Skrimshire (London and New York: Continuum, 2010), 68–9; Chris Abbott, Paul Rogers and John Sloboda, *Global Responses to Global Threats, Sustainable Security for the 21st Century* (Oxford: Oxford Research Group, 2007), 28; Parenti, *Tropics*, 14–20.
38. Tim Flannery, *The Weather Makers: The History and Future Impact of Climate Change* (London: Penguin, 2005), 100–3; Moki Kokoris, 'Perspectives on the Evidence and Impacts of Changing Environments in the Far North', The Arctic Institute, Center for Circumpolar Security Studies, 16 February 2012. http://www.thearcticinstitute.org/2012/02/perspectives-on-evidence-and-impacts-of.html (accessed 11 November 2013).
39. See Hugh Brody, *The Other Side of Eden: Hunter-Gatherers, Farmers and the Shaping of the World* (London: Faber and Faber, 2001), for tour-de-force analysis on the distinction between the mind set of hunter-gatherers, and the rest.
40. Vandana Shiva, *Soil not Oil: Climate Change, Peak Oil and Food Insecurity* (London: Zed Books, 2008), 3, 121.
41. Ibid., 12; Oxfam report, 'He died in front of us. The head of the family died in front of his wife and children can you imagine?', as cited in Alex Renton, 'India's Hidden Climate Change Catastrophe', *Independent on Sunday*, 2 January 2011. http://www.independent.co.uk/environment/climate-change/indias-hidden-climate-change-catastrophe-2173995.html.
42. Helena Paul and Ricarda Steinbrecher, *Hungry Corporations: Transnational Biotech Companies Colonise the Food Chain* (London and New York: Zed Books, 2003), esp. Chapter 8, 'Opening Up the South'.
43. Renton, 'India's Hidden Climate Change Catastrophe'.
44. Fred Pearce, 'King Cotton: Drought, Diggers and Dirty Secrets' in *Confessions of an Eco-Sinner: Travels to Find Where My Stuff Comes From* (London: Eden Project Books, 2008), Chapter 11.
45. The scale of the April 2013 Savar building collapse in Dhaka, Bangladesh in which over 1100 mostly garment-workers were killed, and the entirely ephemeral spotlight it placed on the co-responsibility of major Western clothing brands, actually underscores the degree of First World indifference to Third World suffering.
46. Biofuelwatch. http://www.biofuelwatch.org.uk (accessed 11 November 2013), for regular updated information.
47. Shiva, *Soil*, 89–92.
48. Even then, the major Naxalite attack in Chhattisgarh in late May 2013 on a Congress Party convoy leaving at least 27 dead, leading politicians included, hardly registered in Western news coverage; see http://en.wikipedia.org/wiki/2013_Naxal_attack_in_Darbha_valley; 'Chhattisgarh Naxal Attack/24 Killed'. http://www.ndtv.com/blog/show/chhattisgarh-naxal-attack-24-killed-371356 (accessed 11 November 2013).
49. René Dumont, 'Agriculture as Use and Transformation of Nature', in *Peasants and Peasant Societies*, ed. Teodor Shanin, 2nd ed. (Oxford and New York: Basil Blackwell, 1987), 151.
50. See John Berger, *Pig Earth* (London: Vintage Books, 1992), for a magnificent portrayal.
51. Joachim Radkau, *Nature and Power: A Global History of the Environment*, trans. Thomas Dunlap (Washington, DC: German Historical Institute; Cambridge: Cambridge University Press, 2008), for a masterful overview; also Paul Martin, 'The Discovery of America',

Science 179 (1973): 969–73, for a key exposition of the argument for a major wipe-out of a range of fauna by hunter-gatherer societies at the onset of the Holocene.
52. See Jared Diamond, *Collapse: How Societies Choose to Fail or Survive* (London: Penguin, 2005); Joseph A. Tainter, *The Collapse of Complex Societies* (Cambridge: Cambridge University Press, 1988).
53. Fermand Braudel, *Civilisation and Capitalism: 15th–18th Century*, 3 Volumes, trans. Sian Reynold (Berkeley, CA: University of California Press, 1979).
54. Shiva, *Soil*, 13–14.
55. Ibid., 110–15.
56. Nicola Peel, *Blood of the Amazon* (2011). http://www.eyesofgaia.com./pb/wp_e91307a5/wp_e91307a5.html (accessed 11 November 2013).
57. Ibid.; ChevronToxico, 'The Campaign for Justice in Ecuador'. http://chevrontoxico.com/about/health-impacts/ (accessed 11 November 2013).
58. See Huseman and Short, 'A Slow Industrial Genocide', esp. 220–1, 224–7, for the parallel and strikingly similar Canadian experience of chronic 'war zone' pollution from massive quantities of toxic waste water required to process oil from tar sand deposits in the once bio-diverse lower Athabasca river region, with resulting 'biological warfare' on its native peoples, plus both federal and provincial government efforts to deny the medical evidence of the deadly carcinogenic results. For a highly informative and lucid discussion of the genocidal impact of tar sands extraction, see also Martin Crook and Damien Short, 'Marx, Lemkin and the Genocide–Ecocide Nexus', *The International Journal of Human Rights* 18, no. 3 (2014): 298–319.
59. Marisa Handler, 'The New Amazon'. http://www.orionmagazine.org/index.php/articles/article/10/ (accessed 11 November 2013).
60. chevrontoxico.com/about/historic-trial/about-the-trial.html (accessed 11 November 2013).
61. Peel, *Blood.*
62. Parenti, *Tropic*, 47.
63. Deborah MacKenzie, 'The End of Civilisation', and 'Are We Doomed?', *New Scientist*, 5 April 2008.
64. Levene, 'The Apocalyptic as Contemporary Dialectic', 59–61.
65. Kazimierz Dobrowolski, 'Peasant Traditional Culture', in *Peasants and Peasant Societies*, ed. Teodor Shanin, 2nd ed. (Oxford and New York: Basil Blackwell, 1987), 261–77; and John Berger, 'The Vision of a Peasant', in *Peasants and Peasant Societies*, ed. Teodor Shanin, 2nd ed. (Oxford and New York: Basil Blackwell, 1987), 278–83.
66. Jared Diamond, *The World Until Yesterday: What Can We Learn From Traditional Societies?* (London: Allen Lane, 2013), for a recent, ostensible incarnation of the argument. Even so, Diamond's environment-cum-geography thesis is notable for its absence of engagement with traditional cultures' 'realm of ideas'. See Wade Davis' review in *The Guardian*, 9 January 2013. Much closer to the argument herein, see Alastair McIntosh, *Hell and High Water, Climate Change, Hope and the Human Condition* (Edinburgh: Birlinn, 2008), 'Part 2 – The Human Condition'.
67. See Mark Levene, *Genocide in the Age of the Nation State*, 2 Volumes (London and New York: I.B. Tauris, 2005).
68. Patrick Wolfe, 'Structure and Event, Settler Colonialism, Time and the Question of Genocide', in *Empire, Colony, Genocide, Conquest, Occupation and Subaltern Resistance in World History*, ed. Dirk Moses (Oxford and New York: Berghahn Books, 2008), 119–22.
69. Shiva, *Soil*, 131.
70. Ibid., 143.
71. Parenti, *Tropic*, 208.
72. Ibid., 209; Levene, 'The Apocalyptic as Contemporary Dialectic', 61–3.
73. See Dyer, *Climate Wars*, for the geopolitics and some truly terrifying future scenarios.
74. Lewis Williams with Rose Roberts and Alastair McIntosh, 'Introduction: Human Ecology: A Pedagogy of Hope', in *Radical Human Ecology, Intercultural and Indigenous Approaches*, eds. Lewis Williams, Rose Roberts and Alastair Macintosh (Farnham and Burlington, VT: Ashgate, 2012), 4.

Marx, Lemkin and the genocide–ecocide nexus

Martin Crook and Damien Short

Human Rights Consortium, School of Advanced Study, University of London, UK

A number of studies have shown that ecocide can be a method of genocide if, for example, environmental destruction results in conditions of life that fundamentally threaten a social group's cultural and/or physical existence.[1] With the ever-increasing rise of such cases of ecological destruction brought on by the extractive industries, or indirectly induced by anthropogenic climate change, we argue that the field of genocide studies should draw from the rich scholarly tradition of *political ecology* and *environmental sociology*. Indeed, it is the contention of the authors that, given the looming threat of runaway climate change in the twenty-first century, the advent of the geological phase classified by geologists and earth scientists as anthropocene[2] and the attendant rapid extinction of species, destruction of habitats, ecological collapse and the self-evident dependency of the human race on our biosphere, ecocide (both 'natural' and 'manmade') will become a primary driver of genocide. It is therefore incumbent upon genocide scholars to attempt a paradigm shift in the greatest traditions of science[3] and to cohere a synthesis of the sociology of genocide and environmental sociology into a theoretical apparatus that can illuminate the links between, and uncover the drivers of, ecocide and genocidal social death.[4] Following a discussion of both the conceptual and legal nexus between ecocide and genocide, we further contend that capitalist 'land grabs' carried out by extractive industries, industrial farms and the like are, through the annexation of indigenous land and the associated 'externalities', the principal vectors of ecologically induced genocide when the *genos* in question is an indigenous people.

'Ecological' genocide and the capitalist instrument

Ecology is the scientific study of organisms, the communities they collectively comprise and their interactions with their abiotic environment: *ecosystems*. Political ecology has many definitions[5] but the one with greatest utility for the purposes of this article is that furnished by Blaikie and Brookfield:

> The phrase 'political ecology' combines the concerns of ecology and a broadly defined political economy. Together this encompasses the constantly shifting dialectic between society and land-based resources, and also within classes and groups within society itself.[6]

In other words, political ecology is the study of the community of organisms and their environments that critically includes humans or social collectives. Consistent with an ontological approach that argues it is not necessarily the intention of any particular actor that is to blame for genocide but the effect of what Barta calls 'remorseless pressures of destruction inherent in the very nature of the society'[7] that is of importance,[8] the argument put forward in the following section of the article is that one can best explain the driving force behind ecologically induced genocide by examining the structural forces underpinning the aforementioned industrial mining and farming industries: namely the political economy of capitalism and the capitalist mode of production (MOP).[9] Indeed, it will be argued that only by employing Karl Marx's critique of the political economy of capitalism combined with the sadly overlooked theory of ecology found principally in *Capital* Vols 1 and 3, the *Grundrisse* and *Theories of Surplus Value* can the drivers behind ecological destruction as a genocidal technique be explained. Ecocide will be understood as a function of capital, with its remorseless drive to accumulate damaging and collapsing natural cycles and turning them into 'broken linear processes',[10] exceeding the constraints and boundaries of nature and causing what Marx described as a 'metabolic rift' between humankind and nature. Together with Marx's critique of capital and in particular his analysis of the value form under generalised commodity production for the market, Marx's method serves as an invaluable tool kit to critique ecocide and its necessary corollary: genocide.[11] Furthermore, the drivers in the capitalist MOP that lead to ecological destruction are the very same drivers that have triggered climate change and the anthropocene. The ceaseless drive to accumulate capital and the resultant environmental degradation (what Marx called the metabolic rift) are subjecting the biosphere itself to ecological stresses that could cause the ecological collapse of human civilisation and perhaps trigger an auto-species extinction event.[12]

Marx's ecology

As early as the publication of *The German Ideology*, Marx recognised the co-evolution of human and natural history:

> We know only one science, the science of history. History can be viewed from two sides: it can be divided into the history of nature and that of man. The two sides, however, are not to be seen as independent entities. As long as man has existed, nature and man have affected each other.[13]

Marx analyses human history from the standpoint of the *social* production of wealth or what he calls *use values*, which includes basic requirements of food, shelter and clothing, as well as cultural and aesthetic needs:

> The first premise of all human history is, of course, the existence of living human individuals. Thus the first fact to be established is the physical organisation of these individuals and their consequent *relation to the rest of nature* (emphasis added).[14]

It is clear from this and other passages that Marx sought to develop a theory of society and history that was materialist on both a *social* and *ecological* level. Therefore, when seeking to uncover the social forces that drive and shape society, the social facts that must be established are both the prevailing material conditions in the sense of what Marx called the forces of production – labour power, the tools, infrastructure, instruments and techniques inherited form the past – but also the *natural conditions* that must be appropriated to reproduce labour itself[15] and provide the natural resources and raw materials in the social production of wealth.

These natural conditions Marx defined as a form of *primary* appropriation of use values, use values spontaneously given by nature.[16] Once these social facts have been empirically verified, one can infer the corresponding social relations. Marx understood, as ecologists today do, that natural systems have a metabolism that governs the exchange of matter and energy. The primary focus of the science of ecology is illuminating the nexus of processes (such as primary production, pedogenesis or soil formation, nutrient cycling, etc.) responsible for regulating the exchange of energy and matter (metabolism) in ecosystems. These processes allow for the regeneration and continuation of ecosystems and the communities of organisms they contain.

Marx averred that humankind depends on these natural metabolic processes for its reproduction and continuance as well and so he extended the ecological focus to include human societies in the web of ecosystems. The critical lynchpin in our nexus with the ecosystems that act as our natural incubators is, Marx identified, *labour.*[17] Labour is the principal mechanism that regulates humankind's metabolic relationship with the rest of nature: it is labour that mediates the exchange of materials and energy between humans and nature, but in turn this *metabolic* interaction is itself socially mediated by the historically structured social relations between producers and between producers and appropriators of the surplus product.[18] In other words, Marx recognised that human production or appropriation of wealth is mutually constituted by social relations *and* the contents and constraints of natural conditions.[19] So, although there are what Marx called 'nature-imposed conditions' that regulate the metabolic process that supports all life on earth, they are also affected by human agency according to the particular forms of social interaction or social relations that were imposed on nature.[20] Marx's methodology is neither *anthropocentric* nor is it *nature-centred*, rather it is a dialectical unity of the two.[21]

The metabolic rift

Thus, Marx ecologically embeds all societies. The social metabolic order imposed by capitalism, however, according to Marx, leads to a 'metabolic rift' and the forcible historical divergence of human production from the evolving natural world precisely because of the form of social mediation of the metabolic process under the capitalist economy.[22]

It is here that Marx's critique of political economy illuminates the mechanisms at play explaining the rift in the social metabolic order. Capitalism's central motor force is the drive to accumulate capital and, thus, in the process, all other social and natural relationships are subordinated to this primary goal. As Paul Sweezey argued, it is 'both the subjective goal and the motor force of the entire economic system'.[23] Once generalised commodity production is established, capital accumulation becomes, through the force of competition, a ceaseless and remorseless process, a kind of 'treadmill of accumulation' that respects no other cycles other than the business cycle.[24] Consequently, the insatiable drive to accumulate capital and thus reap profits tramples all over natural cycles and processes and is no respecter of the natural rhythms of regeneration and recycling, as this imperative to accumulate on an ever expanding scale requires more and more of what ecologists term 'throughput of materials and energy'.[25]

The iron law of exponential growth under capitalism exacerbates the social metabolism of the capitalist system and places an ever greater strain on nature, eventually leading to metabolic rift. One feature of a system of universal commodity production for the market is that exchange values regulate the production of social wealth and validate particular labours as what Marx described as 'socially necessary labour time'. In other words, the intrinsic 'value' of a commodity, at least according to capitalism, is the amount of socially average labour that must go into its production. Critically, a market economy based on the organising principle of exchange value presupposes that all producers cannot reproduce themselves independently of the market nexus and thus it assumes the social separation from the natural conditions of production.[26] The social separation of the producers from the natural conditions of production makes possible a compulsion to perform surplus labour beyond immediate needs and ignore natural limits. It also rather palpably demonstrates one of the many ways that capitalism affects our alienation from nature and violates the nature-imposed conditions of social metabolism.

Marx first developed his concept of the metabolic rift in the midst of the crisis of soil erosion in England in the nineteenth century.[27] Marx understood that soil requires nutrients such as phosphorous, nitrogen and potassium to facilitate the growth of crops. But, due to the prioritisation of accumulation of capital and the subsumption of all natural and social relations, the over-exploitation of the soil led to its terminal erosion, triggering an agricultural crisis. This crisis exemplified the tendency of capitalism to violate natural limits and natural rhythms and fundamentally violate the nature-imposed conditions of social production. One marked feature of this transgression was the rise and exacerbation of the division between town and country, first begun by the enclosure movement, itself a function of the need for capital to not only alienate the producers from their natural conditions of production and thus secure the pre-eminence of exchange value, but also to create a home market for its goods.[28]

This led to the ecologically disruptive concentration of the population in ever-increasing urban centres and the movement of large quantities of food and fibre into the cities. Consequently, nutrients that would normally be recycled back into the soil were now accumulating as waste in the cities. It was this kind of ceaseless production of vendible exchange values that led to the severing of the metabolic cord between human civilisation and nature. The industrial sprawls themselves would be damaging to the local environment and health of the urban population and in turn would supply agriculture with the technical means via industrial agriculture to further increase the material throughput and inflict still more damage on the local ecosystems. In the limitless appropriation of the social and natural environment for the exploitation of labour in the form of growing industrial throughput and materially dense populations, capitalism threatened to break down the ecological and biospheric web in its corrosive transmission of matter and energy between and within town and country.[29]

This analysis of environmental crisis under capitalism has the potential to be extrapolated on the *macro* or *global* biospheric scale and explain the advent of *ecological imperialism.* Marx continually stressed that capital, in its remorseless pursuit of value, would span the globe.[30] This geographic expansion is a natural function of the rise of the world market, the international division of labour and the drive to expand industrial production in the pursuit of capital accumulation. With this analysis it is possible to envisage the division of the world between industrialised nations and those countries and regions, such as the territories of the indigenous peoples in northern Alberta, Canada, which supply materials and resources, shrouded in an ecological pall. Moreover, it is possible to explain the advent of climate change as the expression of a political economy that is structurally compelled to transgress what the Club of Rome called in their report *The Limits to Growth*[31] or what Marx called the natural limits of production.[32]

Now we can turn to Marx's analysis of the value form under capitalism to explain these structural features and why, as we shall see in the final section of this article, they lead to the sorts of episodes of ecocide and genocide currently being experienced by the indigenous peoples of northern Alberta and of the Northern Territory in Australia.[33]

The anti-ecological reign of exchange value

As alluded to already, under capitalism a distinction is made within the category of 'value' between the *use value* of an article, the actual utility or usefulness it has for humanity, and the *exchange value*, which is a signifier for abstract social labour time of an article, which equates to the socially necessary labour time that was invested in the making of the product. It is the latter that is exchanged for money in the process of accumulation, what Burkett calls vendible commodities that 'objectify' abstract social labour.[34]

This has a number of ecological implications. First, as wealth, or value, is made up of both use value and exchange value, capitalism undervalues the contribution of nature to the social production of wealth. As Levene and Conversi argue:

> The now-accelerated drive towards the commodification of everything within a global, capitalist, political economy, has no truck or time for ecological versatility except in so far as it can be quantified in terms of monetary value.[35]

Wealth is in reality made up of 'myriad use values produced by materially variegated forms of labor and nature' and, thus, the contribution of nature to the production of wealth is rendered invisible by the formal abstraction of abstract social labour time.[36] Thus, the political economists before Marx did not acknowledge the vital contribution nature makes to the production of wealth.[37] Burkett succinctly summarized the value–nature contradiction:

> Capitalism only validates human and extra-human nature as necessary parts of human production insofar as they can be profitably objectified in vendible use values.[38]

Second, this obliteration of qualitative differences through the aforementioned formal abstraction in commodities for the purposes of exchange also 'abstracts' from the complex, delicate and intricate web of ecological interconnections and diversities, to the extent that these qualities are not captured by the category of social labour time. When you valorise a commodity 'its natural properties are extinguished; it no longer takes up a special, qualitative relationship towards the other commodities'.[39] The formal abstraction under exchange value therefore tends towards the *simplification* and *homogenisation* of nature as well as its artificial *divisibility* or *fragmentation* into either elements of the natural conditions of production or as commodities themselves: 'as a value, every commodity is equally divisible; in its natural existence this is not the case'.[40] Climate change and the excessive emissions of greenhouse gases into the atmosphere are therefore understood as the violation of the ecological principle of sustainability and the intricate interconnectivity of the various parts of ecosystems and the biosphere more generally. In other words, capitalism is blind-sided by the tendency to fragment, homogenise and divide nature and is thus incapable of grasping the drivers of ecocide and climate change or structurally accommodating systemic solutions and remedies to ecological meltdown and global warming.

Extreme energy: a nadir of the anti-ecological dysfunction of exchange value

Thus, we see how Marx's critique of political economy, and his value analysis more specifically, can help explain the ecologically destructive forces unleashed by the capitalist extractive and farming industries. The capitalist MOP implicated in said industries is structurally geared towards the social production of commodities in a manner in harmony not with nature's laws of conservation, sustainability and the natural metabolic cycles, but in step with the imperatives of capital accumulation and exchange value. Thus, the Athabasca 'tar sands'[41] stands as an example of the artificial division and fragmentation of the local ecosystem in an attempt to extract oil with no regard for the anti-ecological effects this unnatural throughput and transfer of energy and materials has on the local environment and, critically, the local population.

Of course, this value analysis is perfectly capable of accounting for all forms of what have become known as 'extreme energy',[42] a particularly virulent form of ecologically unsound industrial energy extraction. These include, for instance, mountain-top removal, deep-water drilling and hydraulic 'fracking'. But what exactly constitutes or qualifies as a form of extreme energy? A commonly held definition is a 'more intensive and environmentally destructive method of energy extraction than conventional sources of fossil fuels'.[43] But, as Lloyd-Davies argues, this definition still leaves nebulous the issue of just how 'extreme' a form of energy has to be to qualify for inclusion in this category; thus, the definition is light on predictive or explanatory power. Instead, an understanding of extreme energy *as a process* is proffered, where resources that are easiest to extract are targeted first. Once depleted, resources increasingly more difficult to extract follow, resulting in an ever more complex and energy-intensive extraction effort.[44]

The extreme energy process can be explained by Marx's political economic analysis. As explained above, one of the central ecological contradictions of the capitalist MOP is that between the exponential increase in the throughput of materials and energy associated with the treadmill of accumulation and the natural limits of production. Given the disequilibrium between capital's ferocious pace in the throughput of energy and materials and nature's laws and temporal rhythms and metabolic cycles, eventually capital provokes what Marx described as 'materials-supplies disturbances'. This results in an inevitable shortage of materials and an accumulation crisis.[45]

The result, as dictated by the operation of the law of value, is that the price of the relevant raw material will go up as the amount of socially necessary labour time objectified in each individual product or use value rises in relative terms. Marx, analysing this phenomenon through the prism of an agricultural crisis explained:

> a *crisis* can arise: 1. in the course of the *reconversion* [of money] *into productive capital*; 2. through *changes in the value* of the elements of productive capital, particularly of *raw material*, for example when there is a decrease in the quantity of cotton harvested. Its *value* will thus rise.[46]

This rise in the value of constant capital, as opposed to labour, could become so costly it starts to disrupt the process of the reproduction of capital, as the profit realised in the sale of a whole plethora of commodities, of which the various raw materials are a constituent part, no longer covers the costs of the elements of production.

This process is exemplified by extreme energy as the supply of fossil fuels begins to run up against natural limits, thus raising the relative amount of objectified labour in a given quantity of fossil fuel leading, in the medium to long term, to a rise in the average price of fossil fuels. Indeed, within the process of extreme energy, where more complex and costly techniques are required for the extraction of ever-scarcer sources, the very same process unfolds. This is precisely what Marx spoke of in his analysis of materials-supply disturbances as a form of accumulation crisis. Furthermore, Marx argued that the resulting rise in the price of raw materials engenders, under conditions of competitive accumulation, a number of competitive responses. These include increased production from suppliers (therefore accelerating and intensifying the metabolic strain on the environment and exacerbating the aforementioned contradiction) and the use *of previously unused substitutes.*[47] So, extreme energy 'as a process' can be seen as both an expression of material shortages engendered by the contradiction between what Burkett calls nature's time and capital's time and a competitive response, through the operations of the market, to correct the imbalance through the extraction of ever more extreme substitutes. The net metabolic effect on the social metabolism is to put further pressure both on local ecosystems and the biosphere more generally.

One remarkable feature of extreme energy is the amount of energy needed to obtain the scarce resource in the first place, with a marked drop in what is known as the 'energy return on investment' (EROI).[48] Thus, as increasingly more difficult to extract resources are sought, as predicted by the theory of extreme energy as a process, the less net energy is acquired. This manifestly irrational feature of extreme energy is also explained by Marx's value analysis; the only value that really counts in a capitalist economy is exchange value or abstract labour value, with nature, and any regard to the ecological imperatives of sustainability, conservation and energy efficiency not being considered, with one qualifier being, that the latter is considered only in terms of reducing costs of production in order to gain a competitive advantage and thus *sell even more*. The end aggregate result, when all capitals are taken together, is still a process that demands ever-expanding production of saleable use values with its attendant exponential increase in material throughput.

Marx's value analysis and the reign of exchange value under conditions of competitive accumulation, and his ecological concept of the metabolic rift, illuminate both the mechanisms that lead to ecological destruction under capitalism and the consequences on humanity's ecological relationship with the planet. In the case of the former, the hegemony of exchange value renders invisible nature's contribution to the social production of use value. Exchange value also eliminates all qualitative differences in nature and abstracts from the complex and intricate web of connections that make up ecosystems. Furthermore the quantitative formal abstraction of socially necessary labour time tends towards the homogenisation of nature and facilitates its artificial division and fragmentation in the form of discrete commodities or factors of production, with all the devastating ecological consequences this implies.

This sets in motion a rampant process of accumulation, carving up nature and increasing the material throughput of production to ecologically unsustainable levels, disturbing the social metabolism of human civilisation and leading to a metabolic rift of man from nature. One extreme and virulent expression of this rift is, of course, global warming. The extractive industries in general and extreme energy more specifically, are manifestations of the anti-ecological imperatives of capital accumulation of value. Extreme energy as a process is one particularly virulent expression of this metabolic rift and the anti-ecological nature of the value–nature contradiction. The resort to more costly and more environmentally destructive forms of energy extraction exemplified by extreme energy signifies a particular form of environmental crisis under capitalism caused by material shortages and the natural limits of production.

Thus, we have illuminated the nexus between the capitalist MOP and ecological destruction. What remains to be explained is how this nexus can be seen as a structural technique of genocide. To this end, it is necessary to outline an understanding of genocide and, second, to discuss ecocide as a key 'method' of genocide. It is to this task that we now turn.

Genocide and ecocide: the legal and conceptual nexus

In 1933, the Polish jurist, Raphael Lemkin, spoke at the International Conference for Unification of Criminal Law in Madrid. Invoking the linked concepts of 'barbarity' and 'vandalism', Lemkin urged the international community to ban both the physical and cultural destruction of human social groups. In his subsequent seminal text *Axis Rule in Occupied Europe*, Lemkin combined barbarity and vandalism, to form a new master concept, *genocide*, with the Greek word *genos* meaning tribe or race and the Latin *cide* meaning killing/destruction.[49] In a now famous passage from that text, Lemkin wrote:

> Genocide has two phases: one, destruction of the national pattern of the oppressed group: the other, the imposition of the national pattern of the oppressor. This imposition, in turn, may be made upon the oppressed population which is allowed to remain, or upon the territory alone, after removal of the population and the colonization of the area by the oppressor's own nationals.[50]

The second element of Lemkin's prior formulation, vandalism – the destruction of culture – was thus a *method* or *technique* of group destruction.[51] The central ontological assertion for Lemkin was that culture integrates human societies and is a necessary pre-condition for the realisation of individual material needs. In Lemkin's view, culture is key to collective memory and each unique and distinctive cultural group is a *genos* deserving protection. Thus, for Lemkin, it is culture that animates the *genos* in genocide and not the civilian character of victims as Martin Shaw[52] seems to suggest (which would not distinguish the concept from crimes against humanity) or the arbitrary categorisation of a perpetrator as some scholars argue.[53] For Lemkin, a nation's culture was an essential element of world culture and nations possess a life of their own comparable to the life of the individual:

> The world represents only so much culture and intellectual vigour as are created by its component national groups. The destruction of a nation, therefore, results in the loss of its future contributions to the world.[54]

After *Axis Rule*, Lemkin began research on a comprehensive multivolume 'History of Genocide', covering ancient, medieval and modern time periods. Lemkin's notes for this project outline key 'methods and techniques of genocide', which include

> physical-massacre and mutilation, deprivation of livelihood (starvation, exposure, etc. often by deportation), slavery-exposure to death; biological-separation of families, sterilization, destruction of foetus; cultural-desecration and destruction of cultural symbols (books, objects of art, loot, religious relics, etc.), destruction of cultural leadership, destruction of cultural centres (cities, churches, monasteries, schools, libraries), prohibition of cultural activities or codes of behaviour, forceful conversion, demoralization.[55]

The range of the methods and techniques of genocide are all too often ignored when a conceptual centrality is afforded to *physical* killing. As Dirk Moses argues, the extraordinary implication of such thinking is 'that Lemkin did not properly understand genocide, despite the fact that he invented the term and went to great trouble to explain its meaning. Instead, most scholars presume to instruct Lemkin, retrospectively, about his concept, although they are in fact proposing a different concept, usually mass murder'.[56] In those cases where the centrality of culture to Lemkin's concept is argued, it would be better if such authors simply chose another concept, for example, mass murder or crimes against humanity.

Even so, Lemkin's largely nationalistic understanding of the *genos* has led sociologist Christopher Powell to 'update' his theorising by invoking a somewhat less structurally bound appreciation of the social group by invoking Norbert Elias' notion of a 'social figuration' – a more fluid notion that nonetheless has an identifiable form.[57] However, as philosopher Mohammed Abed, argues, even a more fluid collective object (*genos*) must still display certain features if it is to be logically and ethically susceptible to the harm of genocide.[58] Indeed, being mindful of etymology and the social collectives Lemkin was trying to protect (and their 'future contributions' to the world he spoke of), not just any social figuration is capable of being the victim of genocide. As Abed suggests, its members must consent to a life in common, *its culture must be comprehensive* and its membership should not be easily renounced. Under these conditions, the flourishing of the group's culture and social ethos will have profound and far-reaching effects on the well-being of its individual members, such that the destruction of its cultural and social institutions will eventuate in the individuals suffering the harms and deprivations peculiar to the crime of genocide.[59] Indeed, for Lemkin, the social group (*genos*) 'exists by virtue of its common culture'. Accordingly, during the process of construction of the draft United Nations Convention on Genocide, Lemkin argued that 'Cultural Genocide is the most important part of the Convention'.[60] In his 1958 autobiography *Totally Unofficial Man*, Lemkin subsequently wrote:

> I defended it successfully through two drafts. It meant the destruction of the cultural pattern of a group, such as the language, the traditions, the monuments, archives, libraries, churches. In brief: the shrines of the soul of a nation. But there was not enough support for this idea in the Committee...So with a heavy heart I decided not to press for it.[61]

And so, he agreed to drop the cultural method from the law and with it lost something that, in Lemkin's words, 'was very dear to me'.[62] The removal of this method seems to have contributed to a perverse preoccupation, in legal and scholarly realms, with perpetrator intention rather than the genocidal impacts on victims and to the popular (mis)understanding of the crime of genocide as simply racially motivated mass killing.

In the years following the implementation of the 1948 United Nations Convention on the Prevention and Punishment of the Crime of Genocide (hereafter Genocide Convention) many governments began to voice their concerns about its effectiveness. Genocide was still a reality in many parts of the world and the Genocide Convention seemed to offer little to those groups it was designed to protect. This was, in part, due to the narrow interpretation of what constituted genocide described above and the omission of much of the cultural method in the convention. Concern at the lack of utility eventually led to an extensive United Nations (UN) inquiry into the effectiveness of the Genocide Convention and it is in just such a review that we find the first attempt to criminalise environmental destruction in international law.[63] UN papers demonstrate that members and delegates of several UN institutions, including the Sub-Commission on Prevention of Discrimination and Protection of Minorities,[64] the Legal Committee of the General Assembly and the International Law Commission, discussed, at different times over a 40-year period, how to define and criminalise severe environmental destruction. These institutions met frequently to discuss the elements and issues involved in formulating such an international crime, including the level of intent required for an offence to constitute 'ecocide' or 'severe damage to the environment'.[65]

Early discussions of ecocide were triggered by extreme environmental damage being inflicted on Vietnam, Cambodia and Laos through the use of chemical warfare as part of the US campaign there. Because this was the context in which discussions of ecocide began, and because of the urgency and extremity of the harm being done, early definitions of ecocide tended to be restricted to wartime situations in which intent to cause environmental destruction was present and central. The term ecocide was used as early as 1970, when it was first recorded at the Conference on War and National Responsibility in Washington, where Professor Arthur W. Galston 'proposed a new international agreement to ban ecocide'.[66] Galston argued that if

> [genocide] could be perpetuated against humankind...then an attempt to destroy a natural environment qualified as equally disturbing. Such an atrocity required a similar concept – ecocide, or an attempt to wipe out a specific environment.[67]

In making a link with genocide, Galston was suggesting that environmental destruction can have a genocidal impact and also that the environment can be seen as a victim of ecocide in the same way a social group of people can be seen as victims of genocide. In 1972, at the UN Stockholm Conference on the Human Environment, Mr Olaf Palme, then Prime Minister of Sweden, spoke explicitly in his opening speech of the Vietnam War as an 'ecocide'.[68] The Stockholm Conference focused international attention on environmental issues perhaps for the first time, especially in relation to environmental degradation and trans-boundary pollution. Other Heads of State, including Ms Indira Gandhi from India and the leader of the Chinese delegation, Mr Tang Ke, also denounced the Vietnam War on human and environmental terms.[69] While there was no reference to ecocide in the official outcome document of the Stockholm Conference, the potential for a law criminalising ecocide was widely discussed in the unofficial events running parallel to the official UN Stockholm Conference, including at the 'Folkets Forum' – the People's Summit – where a working group on the Law of Genocide and Ecocide was established.[70] 'Almost every popular movement and group of NGOs addressed the issue. A demonstration with 7,000 participants was held.'[71] Dai Dong, a branch of the International Fellowship of Reconciliation,[72] sponsored a 'Convention on Ecocidal War' (CEW) that took place in Stockholm, Sweden.[73] The CEW brought together many people, including Professor Richard A. Falk, expert on the international law of war crimes and Dr Arthur H. Westing and Dr Egbert L. Pfeiffer who were both biologists and was coordinated by John Lewallen.[74] The CEW called for a UN working group on ecocidal warfare, which would, amongst other matters, seek to define and condemn ecocide as an international crime of war.[75] A draft International Convention on the Crime of Ecocide was prepared for UN consideration by Falk and reproduced in a journal article he published in 1973.[76] It recognised that the Genocide Convention was deficient and that there was a need for another international law that could address ecological crimes. Falk's draft convention, though, primarily envisaged ecocide as a military offence that could be committed in times of war or peace provided the requisite intent was present.

As with the crime of genocide, there was much academic debate over what would constitute the crime and, in particular, whether intent to commit destruction of ecosystems was a necessary element of the crime. John H.E. Fried, a specialist in international law and member of the Lawyers' Committee on Nuclear Policy, believed ecocide to denote 'various measures of devastation and destruction which...aim at damaging or destroying the ecology of geographic areas to the detriment of human life, animal life, and plant life'.[77] Even so, it was recognised by others, such as Falk, that ecocide often occurs simply as a consequence of human economic activity rather than being a result of a predetermined, intended direct attack on the environment. Indeed, even though Falk's draft (1973) Ecocide Convention constructed a primarily military offence he explicitly acknowledged at the outset that 'man has consciously and unconsciously inflicted irreparable damage to the environment in times of war and peace'.[78] While Westing stated that, 'intent may not only be impossible to establish without admission but, I believe, it is essentially irrelevant'.[79]

Even though Falk's draft was never adopted, it was considered by the Sub-Commission on Prevention of Discrimination and Protection of Minorities when it prepared a study (referred to earlier) for the UN's Human Rights Commission regarding the effectiveness of the Genocide Convention. The sub-commission was asked to consider the addition of ecocide as well as the possible reintroduction of the cultural method into the Genocide Convention. Special Rapporteur Mr Nicodème Ruhashyankiko prepared the study, with the final draft published in 1978. At this time, many sub-commission members were supportive of the idea that additional instruments be adopted.[80] Within the sub-commission, Mr Abdelwahab Bouhdiba voiced support for criminalising ecocide: 'any interference with the natural surroundings or environment in which ethnic groups lived was, in effect, a kind of ethnic genocide because such interference could prevent the people involved from following their own traditional way of life'.[81] However, Ruhashyankiko concluded

> from the review of the problem of ecocide regarded as a war crime, in chapter IV of the present study, it follows that the question of ecocide has been placed by States in a context other than that of genocide. The Special Rapporteur believes that an exaggerated extension of the idea of genocide to cases of ecocide which have only a, *very* distant connexion with that idea is liable to prejudice the effectiveness of the Genocide Convention.[82]

Nevertheless, in a follow up to the Ruhashyankiko report, the concept of ecocide surfaced again when the sub-commission considered the same basic issue in 1985 – whether or not to expand the Genocide Convention. This time the Special Rapporteur was Mr Benjamin Whitaker.[83] The report stressed the opinion of the members of the sub-commission who were vocal in their support for the inclusion of a crime of ecocide.[84] Even so, in a non-committal conclusion, Whitaker recommended that 'further consideration should be given to this question'.[85] In subsequent discussions in the sub-commission, once again members spoke out in favour of the creation of a law criminalising ecocide within the Genocide Convention. A draft resolution, prepared for the Commission on Human Rights, submitted as part of the review, recommended that Whitaker expand and deepen the study of the notions of 'cultural genocide', 'ethnocide' and 'ecocide'. In addition, a draft article on cultural genocide had also been prepared,[86] although not adopted. Ultimately, in the sub-commission's final report on its 38th session,[87] it was recommended that Special Rapporteur Whitaker further investigate the expansion of the Genocide Convention to include the cultural and ecocidal methods of genocide and report back in its 40th Session, which did not happen. It is unclear from UN records why the review of the Genocide Convention went no further.[88]

In recent years, the campaign to criminalise ecological destruction in its own right, and as a strict liability offence,[89] has been taken up by Polly Higgins' Eradicating Ecocide network which draws attention to the numerous examples of ecocide and its human consequences worldwide, at a time when preventing further ecological destruction could not be more pressing.[90] Eradicating Ecocide is one of a number of campaigns[91] that highlights the particularly devastating impact environmental destruction has on indigenous peoples who depend on the health of their local environment not only for their own physical well-being but also for their spiritual and cultural health. Indeed, for those indigenous peoples who still retain a cultural attachment to land, environmental destruction can have a genocidal impact.

Indigenous peoples at the ecocide–genocide nexus

For indigenous peoples worldwide the 'logic of elimination'[92] that informed frontier massacres in places like Australia and North America, and the assimilationist agendas that followed, now underpin expansionist land grabs driven by the logic of global capitalism.[93] In numerous colonial contexts in the years after 1945, the traditional forms of colonial terror transformed into a 'genocide machine' as the nature of capitalist domination became less overtly racist and more attuned to corporate imperatives.[94] Nowadays, governments frequently dispossess indigenous groups through industrial mining and farming and also through military operations and national park schemes – all of which routinely take no account of core indigenous rights.[95] It is precisely these industries and their attendant capitalist MOPs that are implicated by the above Marxian value and ecological analysis.[96] As Short has argued in the Australian context, the 2007 Northern Territory National Emergency Response Act (often referred to as the Intervention) in Australia introduced by the government of John Howard contained discriminatory compulsory land acquisition measures that were designed, it seems, with an eye to opening up indigenous lands for mineral exploration and development. Indeed, the Intervention's compulsory acquisition of townships has created a dangerous precedent for other Aboriginal lands.[97] In late 2007, the Howard government signed up to the US-led Global Nuclear Energy Partnership initiative (GNEP),[98] which committed Australia to mine and enrich its uranium, export it to other countries, then re-import the resultant radioactive waste to be stored forever in the Australian desert. Approximately 30% of the world's currently identified uranium reserves are found on Northern Territory (NT) indigenous lands; since 2013 the number of exploration licences for uranium in the NT has doubled, with nearly 80 companies either actively exploring or having applied to explore. With the Aboriginal Land Rights (Northern Territory) Act (ALRA) 1976 amendments and the Intervention's compulsory acquisition measures, indigenous peoples will have no effective means to resist the now 'inevitable' increase in uranium mining in Australia,[99] resulting in yet further culturally genocidal pressures on some indigenous groups. As Aboriginal MP John Ah Kit put it at the start of the Intervention, 'this is about the beginning of the end of Aboriginal culture; it is in some ways genocide'.[100]

Of course, such treatment of indigenous peoples and their lands is nothing new, nor is it something peculiar to the Australian settler colonial context. Up until the end of the frontier era in the late nineteenth century, genocidal processes in North America were largely geared towards, and derived from, expansionist policies opening up Indian land for a seemingly limitless influx of settlers. In the post-frontier period, settlement has unquestionably continued to be a pressing factor; however, following the industrial revolution, the Euro–North American genocidal logic became increasingly focused on the elimination of Indian peoples in order to gain access to their territory for purpose of *resource extraction*. In recent years, the demand for plentiful and 'secure' energy resources has resulted in 'the single largest energy policy shift in North America since ... production peaked in 1971'.[101] As Macdonald Stainsby writes:

> Having failed to pacify Iraq and having engendered new regional opposition in Africa, South America, and the Middle East, the U.S. empire has driven oil prices up to new heights – a trend which will continue into the future. Though peak oil has profound implications for the U.S. dollar and the militarized global economy, these prices have, in the short-term, been masterfully recast as U.S. imperialism's latest and greatest asset: *the creation of massive new oil 'reserves' in a politically friendly region which can feed the U.S. domestic oil market.*[102]

This is referring, namely, to the tar sands[103] in Northern Alberta, Canada,[104] widely considered to be the most destructive industrial project on earth by environmental, human rights and indigenous activists alike.[105] 'Tar sands' is a colloquial term used to describe sands that constitute a naturally occurring mixture of sand, clay, water and bitumen – an exceptionally viscous and dense form of petroleum – which has, since the late nineteenth and early twentieth century, been referred to as 'tar' due to its similar viscosity, odour and colour. Once again, this desired energy resource lies almost entirely within the traditional territories of native North Americans and, as such, is another example of the acute threat to indigenous peoples posed by the process of extreme energy, which brings with it large-scale dispossession and the 'externalities' of pollution and environmental degradation.

Canada initiated oil production in the tar sands in 1967 – 'after decades of research and development that began in the early 1900s'[106] – with Suncor Energy Inc. generating roughly 12,000 barrels per day. Even so, the tar sands were not regarded as a significant player in North America's bid to prolong the life of its petroleum-based economy until 2003 – around the time of the American invasion of Iraq.

Prior to this period, the extremely difficult extraction and production processes of tar sands development was considered too expensive to be economically viable, but with oil prices heading toward $150 per barrel, the tar sands not only became viable but the basis for a shift to American reliance on North American petroleum as a source of fuel,[107] and yet again the lives and lands of native peoples would be sacrificed to the 'needs' of the dominant Euro–North American capitalist society.[108] The tar sands have not only seriously affected indigenous lands but are producing horrendous environmental destruction impacting indigenous physical and cultural health.[109] Indeed, environmental pollution from the tar sands[110] has been linked to high levels of deadly diseases such as leukaemia, lymphoma and colon cancer[111] in indigenous communities.[112] For George Poitras, a Mikisew Cree First Nation member affected by tar sands mining in Fort Chipewyan, Alberta, the battle with industrial mining over land and resources comes down to the fundamental right to exist.

> If we don't have land and we don't have anywhere to carry out our traditional lifestyle, we lose who we are as a people. So if there's no land, then it's equivalent in our estimation to genocide of a people.[113]

And as Chief Roxanne Marcel (Mikisew Cree First Nation) states:

> Our message to both levels of government, to Albertans, to Canadians and to the world who may depend on oil sands for their energy solutions, is that we can no longer be sacrificed.

The nexus between ecocide and genocide as a lived experience, while far from a recent development for many indigenous peoples around the world, will likely become a far more frequent occurrence if we take the warnings of climate scientists seriously. Indeed, the survival of many non-indigenous minorities and discrete cultural groups, and with it their 'future contributions to the world' that Lemkin spoke of, will be threatened by current levels of climate change as they are often the most vulnerable within vulnerable states. Furthermore, while other creatures responded to harsh or varied conditions with biological change over time, humans have generally relied upon their ingenuity to survive by creating innovative ways to live and communicate, and pass down knowledge to children – for the human species, culture is our primary adaptive mechanism.[114] But with the culturally genocidal tendencies of global capitalism and its path of accumulation, we are losing our adaptive edge in the midst of not only a climate crisis but, as Joel Kovel argues, a world-wide 'ecological crisis' generated by and extending deeply into 'an ecologically pathological society' and capitalist economy.[115]

Marx's ecological analysis we have drawn on reveals the structural features of capitalism, in particular its tendency to span the globe and impose a world market and world division of labour, which, in this sense, is best understood as a form of *ecological imperialism*. Capitalism is the first economic structure in human history that not only has the potential to destroy ecosystems and local environments but, through the process of the metabolic rift, to imperil the very biosphere itself and potentially induce forms of *pan-global ecological genocides and auto-genocides.*[116] Anthropogenic climate change is perhaps the most vivid and stark expression of the ruination of the biosphere, and the most devastating and ominous symptom of the metabolic rift.

The process of extreme energy will exacerbate the ecological crisis and, if not halted, will condemn whole human societies and ecosystems to the effects of runaway climate change as known *conventional fossil fuel reserves contain twice* the amount of CO^2 it would take to ensure this outcome. If we are to avoid such a scenario and a potential extinction event for mankind, then, as leading climate scientist James Hansen puts it, 'we must rapidly phase out coal emissions, leave unconventional fossil fuels in the ground, and not go after the last drops of oil and gas. In other words, we must move as quickly as possible to the post-fossil fuel era of clean energies.'[117] Tyndall Centre climate scientist Kevin Anderson concurs, 'the only responsible action with regard to shale gas, or any "new" unconventional fossil fuel, is to keep it in the ground.'[118]

The imbroglio of intent and agency

When invoking the types of arguments above one is often met with the objection: but what of the question of perpetrator genocidal intent? This is of course unsurprising, as the issue has mercilessly dogged the field of genocide studies. Given the fissure in the field conditioned by what Schaller called the constellation of cold war politics and the consequent bowdlerisation of the UN Convention of Genocide, there are differing opinions on how genocidal intent can be established.[119] The proceeding analysis is consistent with an approach and ontology that understands genocide as a process, as a structure or architecture of occupation, not one which can be easily attributed to individuals qua the demands of law or what Tony Barta described as 'black –letter intention'.[120] Accordingly, genocide is seen as a structural process that does not depend on an exterminatory, intending agent such as the state, but is more willing to contemplate the 'social forces extant in all modernizing and colonizing societies that seek to sequester indigenous land and kill its owners if they are resisted'.[121] Thus, intent is invested in the structures themselves.

However, despite this article's 'Post liberal' sensibilities, it does not abandon the proposition that intent can be attributed to individuals entirely. Even though genocidal social death can be produced without specific 'intent to destroy', it can be argued that there is what can be formulated as reasonably foreseeable intent. Whatever the underlying motives, certainly the forcible dispossessions are intentional, the exertion of forcible control over peoples' lives is intentional and the moves to forcibly coerce people off their sacred lands are intentional. Although the resulting physical, cultural and mental harm may be the opposite of the alleged motivation and hence not prima facie intentional as such, in traditional British legal parlance 'foresight and recklessness' as to the consequences of action are 'evidence from which intent may be inferred'.[122]

Furthermore, a Marxist ontology, not to be confused with the dogmatic, mechanical and ossified Stalinist counterpart, has always been sensitive to the role of agency and, thus, does not reduce everything to structure. Therefore, a Marxist method can potentially accommodate cries for corroborating criminal intent and responsibility, of either executives or senior-level managers of corporations complicit in ecological destruction, or senior-level state officials and government ministers who aid and abet said corporations. The dialectical interplay between structure and agency is best encapsulated in this famous aphorism: 'Men make their own history, but they do not make it just as they please in circumstances they choose for themselves; rather they make it in present circumstances, given and inherited.'[123]

But this has left subsequent social scientists precious little to develop and formulate a correct balance between structure and agency. There have been some insightful developments in this regard in the last few years by both Marxist and non-Marxist sociologists and social scientists. Bob Jessop, in a grand distillation of decades of work on the Marxist theory of the state, produced a sophisticated engagement with the structure-agency debate. Jessop argues that just as capital is not a thing, but a social relation, so too is the state. Combining the non-dialectical structuralism of the 'capital-logic school' which argued that the state automatically supports capital accumulation and the reproduction of the MOP with the 'class-theoretical approach' which conceived the state as the ever-changing and ever-contested product and outcome of the balance of social or class forces in a given historical juncture, Jessop sought to recognise both the importance of agency and intention of social actors.

For Jessop, structural constraints are not absolute and do not foreclose choices but act strategically and selectively, privileging certain strategies among actors who seek to transform, change or preserve them.[124] Crucially, social actors act reflexively to push against the structures they operate within, giving the latter a degree of flexibility. The 'strategic relational' approach in essence posits structures – in the case of the capitalist MOP this means the relations of production – as simultaneously setting limits and enabling strategies of social actors. In other words, neither structure nor agency can be reducible to the other. Taking his lead from Roy Bhaskar's realist social ontology and similar in essence to Anthony Gidden's 'structuration theory',[125] Alex Callinicos employed just such a formulation in his defence of the historical materialist method.[126] Thus, one can envisage the controlling minds of corporations operating in Alberta's tar sands or in Australian Northern Territory uranium mining, as social actors with intentions and interests working within structures that privilege certain strategies and actors, whilst excluding others (such as indigenous peoples) and which in turn can be stretched within certain limits.[127]

Conclusion

In this article, we have discussed how Marx's value analysis, and the reign of exchange value under conditions of competitive accumulation, coupled with the ecologically grounded concept of the metabolic rift, illuminate the mechanisms driving ecological destruction under capitalism and the consequences on humanity's ecological relationship with the planet. We have also discussed how the process of extreme energy is a manifestation of the anti-ecological imperatives of capital accumulation of value. Indeed, the process of extreme energy is a particularly virulent expression of the metabolic rift and the anti-ecological nature of the value-nature contradiction. The more costly and environmentally destructive extreme energy production methods signify a particular form of environmental crisis under capitalism caused by material shortages and the natural limits of production.

Following the illumination of the nexus between the capitalist MOP and ecological destruction we then moved on to argue that this nexus can be experienced by some social groups as a structural technique of genocide. Beginning with a run through of Lemkin's cultural understanding of genocide, we showed how the, at the time, ineffectual narrowed-down version of the Genocide Convention ultimately led to UN-level discussions about how best to deal with the growing cultural and ecological threats to the survival of distinct human social groups within the framework of an amended Genocide Convention. As we have seen, it was within such discussions that the concept of ecocide was first used in the UN system. Finally, we moved on to argue that while the Genocide Convention was not ultimately amended, the ecological and cultural threats to discrete social groups and indigenous peoples in particular, posed by the machinery of global capitalism and the process of extreme energy, require us to take seriously their predicament and their description of their situation as genocidal.

Indeed, for those indigenous peoples fighting to retain or regain their lands, they are fighting for their *life as distinct peoples* since, for them, their spirituality and cultural vitality is based in and on and with their lands. If we take this point seriously when this relationship is *forcibly* interrupted and breaks down due to expansionist land grabs driven by global capitalism, the genocide lens becomes appropriate. When indigenous peoples who have a physical, cultural and spiritual connection to their land are *forcibly* dispossessed and estranged from their lands, they invariably experience 'social death', and thus, genocide.[128] Furthermore, when indigenous lands are used by extractive industries the inherent corporate preference for externalising environmental costs can lead to physical, as well as cultural destruction.

What is needed from sociologists in the field of genocide studies is more research into the *context and manner* in which distinct social figurations' cultural and physical health is threatened by expansionist land grabs and ecological destruction driven by global capitalism. Such research should utilise a synthesis of the sociology of genocide and environmental sociology to produce a theoretical apparatus that can illuminate the links between, and uncover the drivers of, ecocide and genocidal social death. Indeed, such an approach should unashamedly utilise the analytical lens of *genocide* as assaults on the 'essential foundations of life of national groups' (and ecosystems are perhaps the most important of all such foundations) is what the concept was designed to highlight and prohibit.[129]

Notes

1. See D. Short, 'Australia: A Continuing Genocide?', *Journal of Genocide Research* 12, nos 1–2 (2010), 45–68; and J. Huseman and D. Short, '"A Slow Industrial Genocide": Tar Sands and the Indigenous Peoples of Northern Alberta', *International Journal of Human Rights* 16, no. 1, (2012): 216–37, at 222.
2. J. Zalasiewicz et al., 'Are We Now Living in the Anthropocene', *GSA Today* 18, no. 2 (2008): 4–8.
3. Thomas. S. Kuhn, *Structure of Scientific Revolutions* (Chicago, IL: Chicago University Press, 2012).

4. Claudia Card, 'Genocide and Social Death', *Hypatia* 18, no. 1 (2003): 63–79.
5. For instance, Bruce. M. Russett, *International Regions and the International System: A Study in Political Ecology* (Chicago: Rand McNally, 1967), vii: *'the relation of organisms or groups of organisms to their environment,* I have attempted to explore some of the relations between political systems and their social and physical environment'; Leslie Anderson, *The Political Ecology of the Modern Peasant: Calculation and Community* (Baltimore, MD: Johns Hopkins University Press, 1994), 6.
6. Piers Blaikie and Harold Brookfield, ed., *Land Degradation and Society* (London: Methuen, 1987).
7. Tony Barta, 'Relations of Genocide: Land and Lives in the Colonization of Australia', in *Genocide and the Modern* Age, 2nd ed., ed. Isidor Wallimann and Michael N. Dobkowski (Syracuse, NY: Syracuse University Press, 2000), 237–52, at 239.
8. Levene and Conversi make an interesting point that the citizens of the West, who are responsible for the lion's share of carbon emissions, may be found guilty of 'intentionalist malice' due to their refusal to desist from their addiction to consumerism. Although a compelling argument, the authors of this article would argue that it is those who effectively control and plan social production in Western societies that are ultimately responsible. Mark Levene and Daniele Conversi, 'Subsistence Societies, Globalisation, Climate Change and Genocide: Discourses of Vulnerability and Resilience', *The International Journal of Human Rights* 18, no. 3 (2014): 281–97.
9. A similar argument linking the structural features of the capitalist MOP and genocide in Martin Crook, 'The Mau Mau Genocide: A Neo-Lemkinian Analysis', *Journal of Human Rights in the Commonwealth 1*, 1 (2013). http://journals.sas.ac.uk/jhrc/article/view/1697/0.
10. Brett Clark and John. B. Foster, 'Marx's Ecology in the 21st Century', *World Review of Political Economy* 1, no. 1 (2010): 142–56.
11. The article is not claiming that the Marx canon is sufficient to explain all the complex ecological and environmental problems in the twenty-first century. The disputation is that, without an appreciation of the structural role of the capitalist MOP in environmental degradation, ecocide cannot be addressed or adequately explained.
12. Levene and Conversi make a similar point in the introduction to their contribution, arguing that 'By contrast, the Anthropocene ... is now having biospheric consequences so profound that before the planet re-balances its own dynamic thermal equilibrium a great extinction event, as has happened at other geological boundaries, cannot be ruled out. This would suggest an appropriate terminology not of genocide but *omnicide*.' Levene and Conversi, 'Subsistence Societies, Globalisation, Climate Change and Genocide', The *International Journal of Human Rights* 18, no. 3 (2014): 281–97.
13. Quoted in Clark and Foster, 'Marx's Ecology in the 21st Century', 144.
14. Karl Marx and Frederick Engels, *The German Ideology* (London: Lawrence & Wishart, 1999), 42. Indeed, Marx repeatedly stressed both nature's and labour's contribution to the production of use values or wealth. For instance, Karl Marx and Frederick Engels, 'Economic Manuscript of 1861–63, Third Chapter', in *Collected Works* (New York: International Publishers, 1988), Vol. 30, 9–346: the creation of use values necessarily involves 'appropriation of the natural world for human needs, whether these needs are needs of production or individual consumption'; Karl Marx and Frederick Engels, *Capital*, Vol. I (New York: International Publishers, 1967), 183–4, 43: 'the worker can create nothing without nature, without the sensuous external world'; 'We see then that labour is not the only source of material wealth, of use-values produced by labour. As William Petty puts it, labour is its father and the earth is its mother.'
15. Marx and Engels, *Capital*, Vol. I, 177: Marx states that labour is 'a process in which man and nature participate'.
16. Ibid., 178.
17. For instance, Karl Marx, *Theories of Surplus Value*, Vol. 3 (Moscow: Progress Publishers, 1971), 34: 'Labour is, first of all, a process between man and nature, a process by which man, through his own actions, mediates, regulates and controls the metabolism between himself and nature. He confronts the materials of nature as a force of nature. He sets in motion the natural forces which belong to his own body, his arms, legs, head and hands, in order to appropriate the materials of nature in a form adapted to his own needs. Through this movement he acts upon external nature and changes it, and in this way he simultaneously changes his own nature.'

18. The latter being a feature of class exploitation. Surplus labour, once realised in the form of saleable commodities under capitalism, is transformed into profit or capital to be invested once more in production.
19. Marx therefore avoided the trap of reducing the determination of human evolution and production solely to historically determined social structures, recognising that human development occurred in and through nature and recognised also that ecological crises are historically determined and bear the mark of particular social forms of production. This can be described as a *dialectical metabolic-materialist* analysis.
20. John Bellamy Foster, 'Marx's Theory of Metabolic Rift: Classical Foundations for Environmental Sociology', *The American Journal of Sociology* 105, no. 2 (1999): 381.
21. Thus, any notions that Marx and Engels advocated 'mastery of nature' are categorically refuted by the above analysis and a passage Engels wrote in Frederick Engels, 'The Part Played by Labour in the Transition from Ape to Man', in *Marx and Engels Collected Works* (Beijing: People's Press, 1972),Vol. 3, 9–346: 'We by no means rule over nature like a conqueror over a foreign people, like someone standing outside nature – but that we, with flesh, blood and brain, belong to nature, and exist in its midst, and that all our mastery of it consists in the fact that we have the advantage over all other creatures of being able to learn its laws and apply them correctly.'
22. Levene and Conversi call the denizens of such societies that have experienced a social metabolic rift as *Homo anthropocenus*. They are 'entirely dependent as a consumer within a globalised, inherently non-sustainable monoculture – one which takes from the global commons but gives nothing back in return and when it finally breaks down, as it *must* under the weight of this *negative equity* [emphasis added]. Levene and Conversi, 'Subsistence Societies, Globalisation, Climate Change and Genocide', *The International Journal of Human Rights* 18, no. 3 (2014): 281–97.
23. Paul Sweezy, 'Capitalism and the Environment', *Monthly Review* 41, no. 2 (1989): 1–10.
24. John Bellamy Foster, 'The Treadmill of Accumulation', *Organization & Environment* 18, no. 1 (2005).
25. See Paul Burkett, *Marx and Nature* (New York: St. Martin's Press, 1999), 112.
26. Ibid., 58.
27. For a detailed exposition of the soil erosion crisis and Marx's analysis see John Bellamy Foster, *Marx's Ecology: Materialism and Nature* (New York: Monthly Review Press), Ch. 5.
28. Marx and Engels, *Capital*, Vol. I, 747.
29. Burkett, *Marx and Nature*, 125, 129.
30. Karl Marx and Frederick Engels, *The Communist Manifesto* (London: Bookmarks Publications, 2010), 12: 'The need of a constantly expanding market for its products chases the bourgeoisie over the whole surface of the globe. It must nestle everywhere, settle everywhere, establish connections everywhere...All old-established national industries have been destroyed or are daily being destroyed. They are dislodged by new industries, whose introduction becomes life and death question for all civilised nations, by industries that no longer work up indigenous raw material, but raw material drawn from the remotest zones.'
31. Donella H. Meadows, Dennis L. Meadows, Jorgen Randers, William W. Behrens III, *The Limits to Growth* (New York: Universe Books, 1972).
32. For the nexus between what Kovel calls the enmity between capital and nature and the broader climate change implications, see Joel Kovel, *The Enemy of Nature: The End of Capitalism or the End of the World?* (London: Zed Books 2007).
33. See Huseman and Short, 'A Slow Industrial Genocide'; and Short, 'Australia'.
34. Burkett, *Marx and Nature.*
35. Levene and Conversi, 'Subsistence Societies, Globalisation, Climate Change and Genocide', *The International Journal of Human Rights* 18, no. 3 (2014): 281–97.
36. Ibid., 81.
37. According to Marx, free appropriation of natural conditions takes place whenever those conditions aid in the production of use value but do not add 'value', in the sense of representing abstract labour time. See Marx and Engels, *Capital*, Vol. I, 204.
38. Burkett, *Marx and Nature*, 83.
39. Karl Marx, *Grundrisse* (New York: Vintage, 1973), 141.
40. Ibid. Marx describes this as the 'dissolving effect of money'.
41. See Huseman and Short, 'A Slow Industrial Genocide', for a full definition of this extraction process.

42. The term was first coined by Michael T. Klare in *The Relentless Pursuit of Extreme Energy: A New Oil Rush Endangers the Gulf of Mexico and the Planet*. http://extremeenergy.org/2013/03/12/the-relentless-pursuit-of-extreme-energy-a-new-oil-rush-endangers-the-gulf-of-mexico-and-the-planet/.
43. E. Lloyd-Davies, '*Extreme Energy*: A Process Not a Category'. http://extremeenergy.org/2013/07/25/defining-extreme-energy-a-process-not-a-category/.
44. Ibid.
45. Marx's analysis of accumulation crisis brought on by materials-supplies disturbances operates on two levels: first, focusing on the conditions of crisis caused by fluctuations in the value of the materials in question brought on by shortages, and the second, relating to the indirect fluctuations in 'prices' brought on by the resultant competition, speculation and the credit system. See Karl Marx, *Theories of Surplus Value*, Vol. 2 (Moscow: Progress Publishers, 1968), 515. For further elaboration of the contradiction between 'nature's time' and 'labour's time' see Karl Marx and Frederick Engels *Capital*, Vol. III (New York: International Publishers, 1967), 118.
46. Marx, *Theories of Surplus Value*, Vol. 2, 515.
47. Marx and Engels, *Capital*, Vol. III, 118–19.
48. Lloyd-Davies, '*Extreme Energy*'.'
49. Raphael Lemkin, *Axis Rule in Occupied Europe: Laws of Occupation – Analysis of Government – Proposals for Redress* (Washington, DC: Carnegie Endowment for International Peace, 1944): 79–95. For further discussion on this see Dirk Moses, 'Raphael Lemkin, Culture, and the Concept of Genocide', in *The Oxford Handbook of Genocide Studies*, ed. Donald Bloxham and A. Dirk Moses (Oxford: Oxford University Press 2010), 19–41.
50. Lemkin, *Axis Rule in Occupied Europe*, 79.
51. See Moses, 'Raphael Lemkin', 33.
52. M. Shaw, *What is Genocide?* (Oxford: Polity, 2007).
53. For more on this point see D. Short, 'Cultural Genocide and Indigenous Peoples: A Sociological Approach', *The International Journal of Human Rights* 14, no. 6 (2010): 836–7.
54. Lemkin, *Axis Rule in Occupied Europe*, 91.
55. M.A. McDonnell and A.D. Moses, 'Raphael Lemkin as Historian of Genocide in the Americas', *Journal of Genocide Research* 7, no. 4 (2005): 504–5. And also in Dominik J. Schaller and Jürgen Zimmerer, *The Origins of Genocide: Raphael Lemkin as a Historian of Mass Violence* (London: Routledge, 2009).
56. Moses, 'Raphael Lemkin', 3.
57. See Christopher Powell, 'What Do Genocides Kill? A Relational Conception of Genocide', *Journal of Genocide Research* 9, no. 4 (2007): 527–47.
58. See M. Abed, 'Clarifying the Concept of Genocide', *Metaphilosophy* 37, nos 3–4 (2006): 308–30.
59. Ibid.
60. Dirk Moses, 'Empire, Colony, Genocide: Keywords and the Philosophy of History', in *Empire, Colony, Genocide: Conquest, Occupation, and Subaltern Resistance in World History*, ed. Dirk Moses (Oxford: Berghahn Books, 2008), 3–54, 12–13.
61. In J. Docker, 'Raphael Lemkin's History of Genocide and Colonialism' (Contribution for United States Holocaust Memorial Museum, Center for Advanced Holocaust Studies, Washington, DC, 26 February 2004), 3. http://www.ushmm.org/genocide/analysis/details/2004-02-26/docker.pdf.
62. Ibid.
63. Sub-Commission on Prevention of Discrimination and Protection of Minorities. Study of the Question of the Prevention and Punishment of the Crime of Genocide. Prepared by Mr Nicodème Ruhashyankiko, 4 July 1978. E/CN.4/Sub.2/416.
64. The Sub-Commission on Prevention of Discrimination and Protection of Minorities undertakes studies and makes recommendations to the commission concerning the prevention of discrimination against racial, religious and linguistic minorities. Composed of 26 experts, the sub-commission meets each year for four weeks. It has working groups and established Special Rapporteurs to assist it with certain tasks. http://www.un.org/rights/dpi1774e.htm (accessed 16 July 2012).
65. In international forums the UN Sub-Commission on Prevention of Discrimination and Protection of Minorities used the term 'ecocide' to describe a potential crime prohibiting environmental destruction, while in later years the International Law Commission preferred narrower formulations based around the notion of 'severe damage to the environment'.

66. *New York Times*, 26 February 1970; quoted in Barry Weisberg, *Ecocide in Indochina* (San Francisco: Canfield Press, 1970).
67. Amy M. Hay, 'University of Texas—Pan-American', H: Environment Roundtable Reviews 2, no. 1 (2012): 15–19. https://www.h-net.org/~environ/roundtables/env-roundtable-2-1.pdf.
68. Tord Björk, *The Emergence of Popular Participation in World Politics: United Nations Conference on Human Environment 1972* (Stockholm: Department of Political Science, University of Stockholm, 1996), 15. http://www.folkrorelser.org/johannesburg/stockholm72.pdf (accessed 16 July 2012).
69. Ibid.
70. Ibid.
71. Ibid.
72. An independent organisation (1970–1976) that built awareness among governments and society on damage to nature by human misuse of technology and chemical products.
73. The purpose of the convention was to describe the destruction of the Indochinese peoples and environments by the United States government; and to call for a United Nations Convention on Ecocidal Warfare, which would receive evidence of the devastation of the human ecology of Indochina caused by the Indochina War, determine which belligerent caused that devastation, request reparations from the responsible belligerent or belligerents and seek to define and proscribe 'Ecocide' as an international crime of war. http://www.aktivism.info/rapporter/ChallengingUN72.pdf (accessed 16 July 2012.)
74. Ibid.
75. Ibid.
76. Richard A. Falk, 'Environmental Warfare and Ecocide – Facts, Appraisal, and Proposals', in *Bulletin of Peace Proposals 1973*, Vol. 1, ed. Marek Thee (Oslo; Bergen; Tromsö: Universitersforlaget, 1973), 80–96.
77. John H.E. Fried, 'War by Ecocide', in *Bulletin of Peace Proposals 1973*, Vol. 1, ed. Marek Thee (Oslo; Bergen; Tromsö: Universitersforlaget, 1973).
78. *Supra* Note 1.
79. Arthur H. Westing, 'Proscription of Ecocide', *Science and Public Affairs*, January 1974.
80. Austria, Holy See, Poland, Romania, Rwanda, Congo and Oman; see E/CN.4/Sub.2/416, 11–117.
81. E/CN.4/Sub.2/SR.658, 53.
82. E/CIT.4/Sub.2/416, 185.
83. E/CN.4/Sub.2/1985/6.
84. E/CN.4/Sub.2/1985/6, para. 33.
85. Ibid.
86. Ibid., 124. Supportive governments: Austria, Holy See, Ecuador, Israel, Oman and Romania.
87. Report of the Sub-Commission on Prevention of Discrimination and Protection of Minorities on its 38th session, Geneva, 5–30 August 1985. E/CN.4/Sub.2/1985/57.
88. The authors are currently researching why the review process stopped here.
89. On this see P. Higgins, D. Short and N. South, 'Protecting the Planet: A Proposal for a Law of Ecocide', *Crime, Law and Social Change* 59, no. 3 (2013): 251–66.
90. This proposition for a the law of ecocide ties in with Marxian ecological analysis even though the preceding analysis may seem out of step with the initial location of ecologically induced genocide in the capitalist MOP *a la* Marx. After all, Marxism is an emancipatory project with revolutionary implications and is not best known for advocating legal and social *reform*. But to conflate reforms won through struggle with the essential and ultimate reform ability of the capitalist system would be a mistake. In fact, the history of struggle for social justice under capitalism attests to the fact that social reforms wrested from the system, although perhaps not always the best long-term solution, can limit and curb the excesses of the capitalist MOP. In particular, Marx's analysis of the legal restraints introduced in nineteenth-century England on work-time in factories is instructive in this case. He argued that the over-extension of the working day, caused by the treadmill of accumulation, was equivalent to the over-exploitation and degradation of the soil. In the former, the recuperative powers of what he called the 'life force of the nation' were threatened and in the latter it was the regenerative powers of the land that were endangered. Thus the ability of capitalism *to reproduce itself* in the long-term was threatened by short-term monetary gain. It is here that he argues that only with *general political* working-class pressure and state interference, and not on the

level of private economic negotiation between labour and capital, what he recognises as a public good – the common societal pool of labour power – can and must be protected – see Karl Marx, *Value, Price and Profit* (New York: International Publishers, 1976), 59. Furthermore, it is in the interest of capital to reproduce itself and ensure its long-term survival, see Marx and Engels *Capital*, Vol. I, 266.

91. For example, Survival International, Forest Peoples' Alliance, Raven Trust, Cultural Survival and many more.
92. P. Wolfe, 'Settler colonialism and the elimination of the native', *Journal of Genocide Research* 8, no. 4 (2006): 387–409.
93. Again Crook offers a similar analysis where the capitalist MOP is identified as the principal mechanism responsible for 'land grabs' and, thus, genocide for territorially bounded peoples. Crook, 'The Mau Mau Genocide'.
94. Davis and Zannis in Dirk Moses, ed., 'Conceptual Blockages and Definitional Dilemmas in the "Racial Century": Genocides of Indigenous Peoples and the Holocaust', *Patterns of Prejudice* 36, no. 4 (2002), 7–36, 24.
95. In particular, the right to 'free prior and informed consent' of those indigenous peoples affected by them – now an established international core principle most recently enshrined in Article 19 of the United Nations Declaration on the Rights of Indigenous Peoples. http://www.un.org/esa/socdev/unpfii/en/drip.html.
96. A full and exhaustive political economic analysis in the vein of a Marxian focus on the totality of a social formation would require an integration of an analysis of the role of the state and other political and ideological forces in dialectically constituting particular historical junctures. Unfortunately this is beyond the scope of this article.
97. D. Short, 2010, *Australia: A Continuing Genocide*?
98. See http://www.gneppartnership.org/.
99. See comments by Resources Minister Martin Ferguson who recently stated 'there's going to be uranium mining on an increasing basis in Western Australia, South Australia and the Northern Territory, we'll see uranium mining in Queensland in due course'. http://www.abc.net.au/news/stories/2009/07/21/2631570.htm.
100. 'Indigenous Intervention "genocide"', AAP, 7 August 2007. http://www.news.com.au/story/0,23599,22202385-29277,00.html.
101. Huseman and Short, 'A Slow Industrial Genocide', 223.
102. M. Stainsby, 'Into a Black Hole: Tar Sands and Oil Production in Western Canada', *Upping the Anti Journal* 5. http://uppingtheanti.org/journal/article/05-into-a-black-hole/ (accessed 28 February 2014). Here, with peak oil and the creation of massive new oil reserves, we see again the operation of the law of value as delineated by Marx. The former being an expression of materials-supplies disturbances and the latter the utilisation *of previously unused substitutes*. See Marx and Engels, *Capital*, Vol. III, 118–19.
103. For a discussion of the term 'tar sands' versus 'oil sands' see Huseman and Short, 'A Slow Industrial Genocide'.
104. 'The recoverable oil reserves in Alberta's tar sands are so bountiful that they vie with oil reserves in Saudi Arabia and Venezuela for top status' (K. Peterson, 'Oil Versus Water: Toxic Water Poses Threat to Alberta's Indigenous Communities', *The Dominion: A Grassroots News Cooperative*, Tar Sands Issue – Issue #48 (2007): 12, 31.
105. The United Nations Environment Program, for example, has identified the tar sands 'as one of the world's top 100 hotspots of environmental degradation', International Boreal Conservation Campaign (IBBC) *Canada's Tar Sands: America's #1 Source of Oil Has Dangerous Global Consequences* (2008). http://www.borealbirds.org/resources/factsheet-ibcc-tarsands.pdf; IBBC homepage, 'Resources' (accessed January 2010), 1.
106. M. Humphries, *North American Oil Sands: History of Development, Prospects for the Future* (Order Code RL34258) (USA: Congressional Research Service, 17 January 2008). http://www.fas.org/sgp/crs/misc/RL34258.pdf (accessed September 2011).
107. E. Black, *America With No Plan for Oil Interruption: Ironically, As Price Per Barrel Drops, American Oil Supply From Canada Imperiled* (The Cutting Edge News, 3 November 2008). http://www.thecuttingedgenews.com/index.php?article=896 (accessed September 2011).
108. Peak oil and the attendant rise in the price of oil is here understood as an expression of the contradiction between the natural limits of production and the treadmill of accumulation, what Marx called materials-supplies shortages, as argued above.

109. Huseman and Short, 'A Slow Industrial Genocide', 216–37.
110. A damning report on waterway pollution, highlighting arsenic amongst other highly toxic substances: Kevin P. Timoney, *A Study of Water and Sediment Quality as Related to Public Health Issues* (Sherwood Park, AB: Treeline Ecological Research, 2007).
111. Kim Petersen, 'Oil Versus Water: Toxic Water Poses Threat to Alberta's Indigenous Communities', *The Dominion*, no. 48 (15 October 2007).
112. For example, the Dene, Cree and Metis communities in Treaty 8 and Treaty 11 Territories.
113. Interviewed in Petersen, 'Oil Versus Water'.
114. B.R. Johnston, 'Human Environmental Rights', in *Human Rights: New Perspectives, New Realities*, ed. A. Pollis and B. Schwab (London: Lynne Reinner Publishers, 2000), 95–113, 96.
115. See J. Kovel, *The Enemy of Nature: The End of Capitalism or the End of the World?*, 2nd ed. (London: Zed Books, 2007), xiii.
116. The term 'auto-genocide' is used because some genocides would arguably be the consequence of conscious and unconscious self-destructive actions within the capitalist system of members of the victim social figurations themselves. The plural 'genocides' is also used here because, despite global capitalism's culturally genocidal tendencies, there are still many thousands of distinct social figurations – 'geni' – around the world today.
117. J. Hansen, *Storms of my Grandchildren: The Truth About the Coming Climate Catastrophe and Our Last Chance to Save Humanity* (London: Bloomsbury, 2007), 289.
118. See Yale Environment, '360 Forum: Just How Safe Is "Fracking" of Natural Gas?', Opinion, 20 June 2011.
119. D. Schaller, 'Colonialism and Genocide – Raphael Lemkin's Concept of Genocide and its Application to European Rule in Africa', *Development Dialogue* 50 (2008), 75–93.
120. T. Barta, 'Three Responses to "Can There Be Genocide Without the Intent to Commit Genocide?"', *Journal of Genocide Research* 10, no. 1 (2008): 111–18.
121. D. Moses, 'Conceptual Blockages', *Patterns of Prejudice* 36, no. 4 (2002): 22.
122. The words 'as such' (interestingly absent from the draft Lemkin penned) in the UN Genocide Convention require groups to be intentionally targeted because of who they are and not for any other reason such as economic gain or self-defence. Given that perpetrators may well have multiple reasons for genocidal action it is not surprising that Helen Fein for one has advocated a more sociologically realistic approach – 'sustained purposeful action': see H. Fein, *Genocide: A Sociological Perspective* (London: Sage, 1993), 24. Under such a formula, intent can also be inferred from action, which is entirely consistent with a long-established principle in British common law – that 'foresight and recklessness are evidence from which intent may be inferred', see J. Wien in *R v. Belfon* (1976) 3 All ER 46.
123. Karl Marx, 'The Eighteenth Brumaire of Louis Bonaparte', in *Marx's 'Eighteenth Brumaire': (Post)modern Interpretations*, ed. M. Cowling and J. Martin (London: Pluto Press, 2002).
124. Bob Jessop, *State Power: A Strategic-Relational Approach* (Cambridge: Polity, 2008), 46.
125. Anthony Giddens, *The Constitution of Society: Outline of the Theory of Structuration* (Cambridge: Polity Press, 1984).
126. The one caveat being that the capitalist MOP and the key role of changes in production are the ultimately determining structure in any given social formation in the last instance, given the underpinning role of economic production to the non-productive relations (the state, religion, judicial relations) in what Marx labelled the superstructure and the accumulative effect of advances in productive forces over large stretches of history. See Chris Harman, *Marxism and History: Two Essays* (London: Bookmarks, 1998), 22–30.
127. This formulation also dialectically posits a role for the other subjects through struggle, namely the indigenous peoples as well as the workers who work for the aforementioned corporations. Thus indigenous peoples are conceived as not just objects but subjects of history.
128. Just such a connection is argued by Levene and Conversi when they aver '...any culturally homogenising and/or economically driven project that might seek to violently disrupt or suffocate traditional, diverse, subsistence practice could also be taken to be genocidal'. Levene and Conversi, 'Subsistence Societies, Globalisation, Climate Change and Genocide: Discourses of Vulnerability and Resilience', *The International Journal of Human Rights* 18, no. 3 (2014): 281–97.
129. Short, 'Cultural Genocide and Indigenous Peoples'.

Crystallisations of the global western state in the era of climate change

Gregory Kent

Department of Social Sciences, University of Roehampton, London, UK

Climate change impacts will change world politics and society. Drawing primarily on Martin Shaw's *Theory of the Global State*[1] – as will be shown, a singularly relevant framework – this article tentatively explores the future structure of world politics and global society in the era of climate change. Climate change impacts will restructure how politics is performed and how social relations develop over the coming decades. Climate-driven conflict and genocide make such change more probable. It is proposed that through humanity's efforts to grapple with the complex challenges climate change will bring, cooperation, interconnection and ultimately integration will be extended significantly between states and societies, not least within the hybrid form of the global Western state itself (see below for definition). The classic formulation of a world government with global juridical powers is unlikely to materialise, because of the all too apparent, near universal attachment to 'national' sovereignty, but the enlargement and deepening of the already historic levels of cooperation seen in the operation of existing regimes of governance, in particular, even greater integration of the core Western state seems likely. In addition, the developing sense of human society on a worldwide scale – a key element of globality which underpins the global Western state's development – is bound, given climate change impacts, to intensify, further consolidate this unique state form. The article makes the case for this relationship between climate change impacts and social and political change by arguing that climate impacts will come increasingly to be understood as violations of human rights, particularly in contexts of war and genocide. The developing trend towards humanitarian response (including intervention) will, in more frequent and intense climate crises, re-crystallise the Western state into ever greater action in response to the above, simultaneously structuring change within it. Ultimately, society based on more fully global characteristics and even more tightly bound governance structures are more likely to address effectively the perils of climate change.

Introduction

From one perspective there is perhaps excessive optimism, dare it be said hubris, underpinning the enquiry in these pages: the consequences of climate change are not knowable. Some consequences are predictable with a high degree of confidence, others remain scientifically and theoretically based predictions political leaders and possibly a majority of publics would see as highly speculative. The focus here is not primarily on these consequences, rather on the events which may follow on their heels. The exploratory nature of the following therefore needs to be acknowledged from the outset.

But climate change, as Kutting argues, is probably 'the most complex problem that global policy-makers have faced in human history'.[2] And are climate change impacts and our response to them the defining moral issue of the twenty-first century – as James Hansen contends[3] – on a par with the structuring into politics of human rights concerns that developed in the latter part of the twentieth century in the aftermath of the Holocaust?[4] Our ground is therefore tentative, but complex, and perhaps of unsurpassed ethical importance. We must also acknowledge the possibility that the survival of this world may be at stake.

Some authors see a technological fix as the way out of climate crisis.[5] This article makes the suggestion that it is more likely, given the history of the development of global governance over the last century, that the effect of climate change impacts and resulting conflicts and genocides transform world politics and society in ways that start to mitigate the harm done to humans and the planet. As John von Neumann argued nearly 60 years ago, climate crisis born of 'extensive human intervention' would 'merge each nation's affairs with those of every other, more thoroughly than the threat of nuclear war or any other war may already have done'.[6] In a real sense, shared vulnerability – even if that exposure is unequal – is, according to Vogler and Imber, to be 'regarded as the ultimate form of interdependence'.[7] However, at the same time, it is increasingly recognised that climate change could add to pressure in vulnerable regions in many states, leading to the disintegration of social and political structures at national or local levels.

The twentieth century saw major developments in war (and genocide) that impacted profoundly on the development of the state. The following is a perspective on state development (and the role of wider 'global governance') contradicting the dominant economics-focused globalisation debate as the main driver of change resulting in undermined state power. This article employs the genocide scholar and social theorist Martin Shaw's *Theory of the Global State* – a theory having singular relevance to climate impacts because of the centrality it accords human rights in a dynamic framework that integrates Western power with the concept of governance in developing globality. Shaw's theory is important in part for the simple realisation that the classic formulation of a world government with global juridical powers is unlikely to materialise because of the all too apparent, near universal attachment to 'national' sovereignty.[8] The key element of globality which underpins the global Western state's development and a central concept in Shaw's sociological work in international relations is bound, given climate change impacts, to intensify, further consolidating this unique state form. This article utilises Shaw's insights to critically re-vision politics and society in an era of climate impacts to make the case for the relationship between climate change impacts and social and political change.

Here, an innovative approach is offered to the threat of climate crisis as a new, unstoppable driver of state/governance development that in turn will provide some, albeit tardy, responses to the challenges created.

This article explores three most significant factors that explain why climate crisis will generate major change in politics and society: First and foremost, climate change, while not exclusively damaging to human well-being and livelihood, will have a profound impact on human beings, specifically on their human rights. Human rights – the basic right to life of huge numbers of people is most obviously in peril in certain climate contexts examined below – are widely considered to be fundamental, inalienable and universal. (Within this context the most likely climate impact scenarios are discussed.) Second, the likely climate-driven impacts of war and genocide have historically been key drivers of the development of state form (and relatedly of 'global governance'). The era of the Holocaust and World War II and the major changes in governance brought about with the establishment of the United Nations (UN) system are perhaps the pre-eminent example of this historical process. (Here, the literature on climate-driven conflict is briefly examined.) Third, the rise of the norm of humanitarian assistance, including intervention, is evoked. Such action, led by the increasingly global Western state, while far from sufficient in numerous ways, does prevent and limit suffering whether resulting from natural disasters such as earthquakes, tsunamis or manmade from wars or genocide, through a system of humanitarian response. This article argues, extending Shaw's framework, that the developing trend towards humanitarian response (including intervention) will, in more frequent and intense climate crises, re-crystallise the Western state into ever greater action in response to the above, simultaneously structuring change within it.

It is important to recognise that 'Global' means more than changed understandings of time and space; more important is the new social meaning that these changes have brought forth: the widespread awareness of worldwide connection of human society – society is now constituted primarily by an inclusive human framework, instead of communities, whether tribal, national or religious – the 'infrastructure of the global is not primarily commercial, technological or communicative, but social, cultural and above all political'.[9] Globality will substantially enrich such commonality and has already changed the meaning of human division. Human society can still be divided and violent and may remain so, but the 'balance and terms of the relationship between commonality and division' has shifted in favour of cooperation and integration. Though 'hardly an end point or final condition', globality is coming into being through the development of the global Western State.[10] In the following section, a key driver of these changes, the global democratic revolution, is discussed in connection with climate-driven human rights issues.[11]

Climate rights-driven transformation

These social and political changes will take place in the context of climate change impacts mainly because such consequences will be felt by ordinary human beings across the globe (with stresses and challenges predicted to be much greater in regions where there is less capacity to cope) in the form of human rights violations – most obviously death and injury, the loss of housing and livelihood, deprivation of basic resources such as water supply and food. A number of writers argue that the disparity in power between those who prefer the status quo and those who stand to lose on the one hand, or that where individuals suffer, on the other, at the hands of the state, is key to what makes these *human rights* issues.[12] More important is the ever-greater focus, both of key human rights non-governmental organisations (NGOs) and advocacy groups such as Human Rights Watch and Amnesty International and international organisations working on social, economic and even cultural rights that have broadened the scope of the rights agenda. The advent of the rights-based approach to development has blurred the distinction between human rights and development, making those deprived of the core features of development, rights holders, and those institutions who are obligated to fulfil the rights holders' rights, duty bearers. How will climate change impact on human rights? There is a spectrum of opinion about precisely this issue; for reasons of space, this article focuses on the key Intergovernmental Panel on Climate Change (IPCC) findings from the most up-to-date Assessment Report (AR4) as the most likely to be taken seriously.

The AR4,[13] and its projected impacts, is an extremely complex set of findings and projections based on extensive and authoritative research. Below, some of the most important findings derived from the synthesis report are summarised.[14] The AR4 was not able to 'assess the likelihood, nor provide a best estimate or an upper bound for sea level rise' because 'understanding of some important effects driving sea level rise is too limited'.[15] Sea ice is projected in both Polar regions to shrink under all scenarios of human response. In late summer, sea ice is projected to disappear 'almost entirely by the latter part of the 21st century', clearly affecting indigenous communities in this region.[16] It was considered '*very likely* that hot extremes, heat waves and heavy precipitation events will become more frequent', potentially affecting the vulnerable, possibly causing death in cases.[17] The report's authors concluded 'it is *likely* that future tropical cyclones (typhoons and hurricanes) will become more intense', with greater peak wind speeds and heavier precipitation.[18] Currently, a single tropical storm can kill hundreds, so it is likely thousands will be killed in future extreme events.[19]

By the 2080s, 'many millions more people than today' could suffer floods annually due to sea level rise according to projections. Worst affected will be populations (and industry) of densely populated, low-lying megadeltas in Africa and Asia with small islands being especially vulnerable.[20] Although not explicitly stated, these people can expect death, injury, displacement resulting from loss of livelihood, housing, food and water sources.

The health of millions is 'projected to be affected' as a result of increased malnutrition, increased incidence of diarrhoea, higher numbers of deaths, diseases and injury resulting from extreme weather events.[21] The impact of climate change on water resources will 'exacerbate current stresses on water resources from population growth and economic and land-use change, including urbanisation' affecting possibly one-sixth of the global population.[22] Increased flooding from a considerable future increase in heavy rainfall events in many regions could adversely affect physical infrastructure and water quality.

By 2020, 75–250 million people in Africa are 'projected to be exposed to increased water stress due to climate change'.[23] The lower adaptive capacity of Africa could mean the cost of adaption to arise in sea level 'could amount to at least 5 to 10 percent of GDP'. It was also concluded that in certain countries, rain-dependent agriculture could be diminished by 'up to 50 percent'. In many countries, agricultural production affecting access to food 'is projected to be severely compromised'.[24]

High population will combine with climate impacts to produce widespread morbidity and mortality due to diarrhoea associated with floods and droughts which are expected to rise in East, South and South-East Asia 'due to projected changes in the hydrological cycle'.[25]

By 2050 in Latin America, temperature rise and associated decreases in soil water 'are projected to lead to gradual replacement of tropical forest by savannah in eastern Amazonia'. Semiarid vegetation will 'tend to be replaced by arid-land vegetation'[26] with adverse impact on food security. Across the continent, population 'at risk of hunger is projected to increase' with medium confidence.[27] Rainfall pattern changes and the disappearance of glaciers 'are projected to significantly affect water availability for human consumption, agriculture and energy generation'.[28]

Sea level rise is expected to worsen inundation, storm surge, erosion and other coastal vulnerabilities, threatening vital infrastructure, settlements and facilities that support the livelihood of small island communities.[29]

Generally, the world will see changed frequencies and intensities of extreme weather events; together with sea level rise they will produce 'adverse effects on natural and human systems'.[30] Unexpected or irreversible impacts could result from man-made climate change but will be dependent on 'the rate and magnitude of such change'.[31]

Polar land ice sheet diminution and/or thermal expansion of seawater 'over very long time scales could imply metres of sea level rise' which would result in 'major changes in coastlines and inundation of low-lying areas, with greatest effects in river deltas and low-lying islands'.[32] But current model projections suggest these developments, if global temperature increase is limited to a range of 1.9–4.6°C (relative to pre-industrial), on a millennial time scale, but rapid sea level rise on century time scales 'cannot be excluded'.[33] Another irreversible impact is likely on species extinction rates, some 20–30% 'at increased risk' for a 1.5–2.5°C (relative to 1980–1999) range, 40–70% for in excess of 3.5°C.[34] These impacts could affect human well-being in many ways. At such levels, are we, as Bill McKibben argues, witnessing 'the end of nature'?[35]

There is widespread recognition that climate change is happening.[36] There is considerable divergence about what impacts will be and when they will occur, but few serious commentators diverge much from the main thrust of what the AR4 projects.[37] A key issue states and societies will have to come to terms with is the determination of what are 'acceptable' impacts, or what events or changes constitute real 'danger', whether to themselves or others. As Schneider and Lane suggest, these questions 'are deeply normative, involving value judgments that must be made by decision makers' rather than scientists.[38] The argument here is that human rights are so entrenched in the global layer of governance that engages most states and a myriad of NGOs that these climate-based violations cannot be ignored. Additionally, human rights drive the growing global social ties that bind individuals in distant domains.

New global democratic drivers

As noted above, a major distinguishing characteristic of the *Theory of the Global State* in relation to globalisation is the idea of global democratic revolution. It is revolution in a new sense, requiring a radical redefinition of the concept of revolution:

> not simply a political rather than a social revolution, or a national and international revolution, but a set of changes in which these terms are themselves recast. The global revolution involves new relations of politics, economics and society, as well as of the national and international aspects of all these relations.[39]

The concept is key to climate-rights driven transformation because it 'rescues globality from the realm of unintended, mechanical change. It asserts the role of conscious human agency in global transformation' and the 'intentional actions of many individual and collective human actors'.[40]

The global revolutions in the former communist bloc countries of Central and Eastern Europe had, from this perspective, worldwide scope 'by breaking down the conditions the national-international form of relations it created the conditions for globality and in a sense intended consequences of the participants'.[41] Revolutions have led to a transformation in state structure such that, 'the political structure of social relations on a world scale has fundamentally changed'.[42] The worldwide influence has not been limited to the former Soviet empire; Western-backed authoritarian regimes have been overthrown in significant numbers. The impact on the remaining left-wing authoritarian states has been significant, if not yet decisive; nonetheless, there persists 'a clear sense that further democratic change is inevitable, even if its timing and mechanisms are far from certain'.[43] There have been more limited revolutions in Africa and Asia. However, ongoing developments in the Arab uprisings across the Middle East and South Africa (MENA) region suggest that while this process is far from over, it is also fraught with difficulty and human suffering. It is worth noting that:

> In 1975 at least 68 percent of countries throughout the world were authoritarian [but] by the end of 1995 only about 26 per cent were authoritarian, all the rest having held some sort of competitive elections and adopted at least formal guarantees of political and civil rights.[44]

The global democratic revolution remains incomplete, which is in part a consequence of the lacklustre response of the institutions and leadership of the global Western state, balancing particular interests in 'maintaining elements of the old relations and forms, and compromise even with the most virulent forms of counterrevolution'.[45] The global nature of the ongoing democratic revolution is, according to Shaw, 'rooted in the extent to which... democracy and human rights have become universal values to which individuals and groups can appeal, if need be over and against national state institutions'.[46] Subjugated groups appeal consistently to the founding documents of universal human rights, but also to the international institutions of 'a Western-led world order as guarantors of those rights'.[47] We will undoubtedly see the rise of new revolutionary reactions to climate-driven human rights concerns; such reactions will carry the development of globality further. The role of human rights and democracy as drivers for development of the emergent global state is a connection explored further below.

War, genocide and the state

While not conclusive, the relationship between climate change impacts and war and genocide cannot be dismissed. Given the importance of war and massive human rights violations to the development of the state, it seems inconceivable that climate impact-related conflict will not further advance the largely unrecognised development of governance in the form of the global Western state. Consequently, this specific, likely impact of climate change could have a decisive influence.

The widespread sense of transformation – loosely correlating to globalisation – is driven by change in political structure (allied to social structural change) resulting from war and genocide in ways that are often passed over. Thus the form and consequent power of the state have been transformed – rather than the cliché of weakening – in globalisation. Historically, intra- and inter-state conflict were critical to the establishment of the boundaries of nation-states. Later, colonial wars (and genocides) of conquest helped to define European states as state-empires; the development of the state after World War II (and genocide) rapidly broke up the old state-empire system; World War II and the Holocaust led to the establishment of the UN system of governance; the formation of new bloc-states, Soviet communist and Western liberal-capitalist – already nascent in the previous period – resulted from the Cold War; the end of the Cold War collapsed the Soviet bloc-state and led to the integration of virtually all of Europe into a more integrated remaining bloc-state: The global Western state, a conglomerate or network of state power.

Shaw describes this emergent global state as 'an integrated authoritative organization of violence which includes a large number of both juridically defined states and international interstate organizations'. The core characteristics of this state derive from its earlier 'bloc' form, but the processes of global transformation have re-made it and that term is no longer satisfactory to describe a new state form. This organisation is not an alliance or an intricate set of alliances but rather a new type of state.[48] It is a form beyond the regime or sets of regimes that constitute 'global governance'. This new form comprises a single military core of power relative to other concentrations that have operated in different forms since World War II, including in Iraq in 1991, Bosnia 1992–1995, Serbia 1999 and more recently in Libya in 2011. Re-crystallisation of this hybrid state form is explored below.

In keeping with his development of state theory built on Weber, Mann and Giddens, as 'bordered power centres defined by relations of violence'[49] military alliances are seen as the 'most defining institutional networks' including 'neutral' states attached to the Western state in different ways. The integration of 'states' into the Western state is established by a complex set of overlapping relations and institutions. The Western state's use of violence is structurally underpinned by ever-greater integration with the UN, an international authority structure with wide global (if contested) legitimacy. In terms of 'juridical states' the global state comprises North America, Western Europe, Japan, Australasia, non-North Atlantic Treaty Organisation (NATO) states such as Finland, Ireland and Switzerland, and many non-Western states in Central and Eastern Europe, Latin America, East Asia and Africa which are so 'closely dependent on and increasingly integrated with the Western state as not to constitute autonomous centres of state power'. This state has borders, which define a potential for war with other states, but it 'operates militarily, politically and juridically beyond its territorial base' – the characteristic that defines its global role; thus, its borders alter depending on the incorporation of 'states' in operable structures.

The UN is the most widely recognised worldwide political organisation with human rights embedded in its foundations. In the post-Cold War period, what Shaw calls this 'authoritative layer of state' has expanded such that it should be seen as 'a global layer' of state power. The global layer of state power has four key features: 'global political power and legitimate force; global legal institutions, including means of enforcement; global economic, environmental and social institutions; and a global ideological framework'.[50] It is not simply an 'elaboration of a loose framework of global "governance", involving a loose "multilateral" agglomeration of nation-states, international institutions and civil society actors'.[51] Shaw notes the paradox of this global layer is its involvement of all centres of state power (and other organisations) – sometimes in a critical or limiting way – and its fundamental dependence upon Western state institutions. But the global layer has relative autonomy and 'simultaneously depends on and transforms the power of the West'.

If climate impacts result in new forms and greater incidences of war and genocide, it is not difficult to imagine that further transformations of the global Western state hybrid form and power will occur. An India more integrated into Western state structures might respond to climate crisis-stricken neighbouring Bangladesh very differently to one remaining outside of them. It is probable that such an India would be driven into ever-greater cooperation and integration as a result of such pressures. Recalling India's 'humanitarian intervention' in the 1971 war and genocide in (today's) Bangladesh, a more global effort seems likely. But will conflict-driven massive human rights abuse result from climate crisis?

Climate-driven war and genocide?

The relationship between climate change and security is considered 'highly speculative' by certain war and peace studies scholars.[52] The decline in violent conflict in recent years identified by Goldstein[53] and others,[54] while slow and uneven, may well confound predictions of climate violence.[55] The IPCC's two most recent reports (2001 and 2007, and from early dissemination of the coming AR5) have not elaborated in substantive detail about the security implications of climate change.[56] There are a number of studies of different time scale, region and conflict type that suggest uncertainty about this particular set of impacts. Most work (including a recent special edition the *Journal of Peace Research* dedicated to this issue) examines climate variability and specific change in precipitation and how these relate to conflict through negative impacts on rain-dependent agriculture.[57]

In recent reviews of the climate conflict literature, Bernauer, Bohmelt and Koubi[58] and Buhaug, Gleditsch and Theisen[59] conclude that although environmental change may, under certain circumstances, increase the risk of violent conflict, the existing evidence indicates that this is not generally the case. Nordas and Gleditsch established that there was little support for the climate–conflict nexus in the academic literature, arguing that further research was needed into the kind of violence to be expected, and greater effort needed to be made in unravelling the causal connections posited between climate change and conflict.[60] Zhang et al.[61] in China and Tol and Wagner[62] in Europe examine the climate–conflict nexus over the very long term. With data for a whole millennium (or two in the case of Zhang et al.), they conclude that war was more frequent in *colder* periods, though the connection is weaker once industrialisation takes place. Possibly the most comprehensive academic inquiry into the issue 'offers only limited support for viewing climate change as an important influence on armed conflict'.[63] A brief summary of these findings is presented thematically below.

Adano et al.[64] found that in two districts in Kenya 'more conflicts and killings take place in wet seasons of relative abundance'. In the same country, Theisen found that 'years following wetter years [are] less safe than drier ones', apparently reversing the expected outcome of global warming. Hendrix and Salehyan find, based on a new database of social conflict in Africa, that rainfall deviations in either direction are associated with conflict, but that violent events are more responsive to heavy rainfall.[65] Regarding the related issue of water scarcity, Benjaminsen et al. declare on the basis of the Mopti region of Mali, central Sahel, that there is 'little evidence supporting the notion that water scarcity and environmental change are important drivers of intercommunal conflicts'.[66] Bernauer and Siegfried's study of the Syr Darya catchment found that interstate military conflict is 'unlikely'.[67] Tir and Stinnett argue that water scarcity increases the risk of military conflict but that in their view institutionalised agreements can offset the risk.[68]

Bergholt and Lujala conclude that natural disasters have an adverse effect on economic growth, but that this does not result in a greater risk of conflict.[69] Slettebak's work supports this obliquely in identifying a growing trend in climate-related natural disasters – in turn supporting the argument of crisis sociologists – that people often unite when faced with adversity.[70]

Against this critique, literature on the effects of sea-level rise and conflict – one of the main ways in which pressures for conflict will be created – is absent. There are studies showing that the El Nino/Southern Oscillation (ENSO) has a significant influence on the onset of civil conflict. Raleigh and Kniveton's study noted that in East Africa rainfall deviations in either direction have been associated with conflict, but they distinguish between civil war, found to be 'more likely in anomalously dry conditions', and non-state conflict, which shows greater connection with wet conditions.[71] However there appears to be a fundamental rift between these credible studies of historic climate and conflict data, which focus on establishing a causal connection between the key variables, and those examining current stresses and tensions and map projected problems and challenges onto these situations. It is important to note that none of the articles mentioned above examine urban conflict or one-sided violence, the main focus of our enquiry.

It has been widely recognised that environmental change has been a major factor in violent, indeed genocidal, episodes in Rwanda and Darfur.[72] But as International Alert (IA), a leading NGO in the conflict prevention sphere, recognises, 'climate change alone does not explain either the outbreak or the extent of the violence in Darfur'; in the Sahel, 16 other states have been dealing with the consequences of climate change, but only Sudan has experienced such a level of violence.[73] Indeed IA is sceptical about the risks; they warn that 'there is a risk of over-stating the conflict dimension in an attempt to persuade a sceptical, even disaffected or merely ill-informed public to support cuts in carbon emissions'.[74] That said, IA has identified 46 countries with significant populations as 'facing a high risk of armed conflict as a knock-on consequence of climate change', with a further 56 countries (1.2 billion people) 'facing a high risk of political instability'.[75]

There is also a more plausible argument in relation to climate impacts on intergroup or intercommunal conflict below state level. Levene cites the example of the Chittagong Hill Tracts in Bangladesh, 'a region where issues of environmental degradation and climate change have *already* interacted with a range of other factors to produce genocidal violence'. According to Levene the,

> particular circumstances of Bangladesh-CHT provide[s] insight into how an interaction between rapid climate change and the vagaries of *political* geography could lead to contours of extreme, including genocidal violence.[76]

The case he makes is convincing, but it also supports the argument of Gleditsch, that climate change will interact with multiple variables. It is clear, as Levene indicates, that Bangladesh's tensions and conflict are associated strongly with a rising population on a declining territory. He suggests that in 'near-desert regions the consequences of the biosphere's response in the form of increasing temperatures alongside reduced rainfall may be the final straw in pushing some of the world's already most marginal and challenged human populations over the edge'.[77]

Support for a climate–conflict nexus has come from unexpected quarters: prominent military thinkers in the US argued that 'climate change acts as a threat multiplier for instability in the most volatile regions of the world',[78] and an IISS Strategic Survey in 2007 recognised that the consequences of climate change would be a significant issue for international security but would only become more generally understood once climate change effects were felt.[79]

IA argue it may be unhelpful to devise 'generic scenarios of how climate change will interact with other variables to increase conflict risk' because context is always critical to how variables actually play out. They trace broad consequences that illustrate risks to agriculture, industry and daily use, on IPCC projections, are set to worsen.[80] They add that limited water supply could create several sources of conflict:

> Conflicting claims to water resources have generally not led to violence between states – the record of settling disputes is largely positive, in fact – but there remains a considerable risk of conflict within states as different groups contend for diminishing water resources. Even where these do not lead to open armed conflict, they may well feed pre-existing tensions and exacerbate the fragility of state institutions. Tensions over water can be also worsened by privatising control of the resource without looking after the rights of the poor.[81]

Ward examines whether anthropogenic climate change will increase the risk of genocide in Africa, finding, 'a number of points of concern for Africa, at least in terms of conflict, if not genocide...the projected climatic changes signify further decreases of rainfall in already arid areas, whilst experiencing an upturn in average temperature, which hasten to exacerbate ecological stress'. In these regions, namely the Sahel and sub-Saharan Africa – traditionally dependent on rainfall agriculture and unfavourable demographic constraints – 'there is real danger of future conflicts and genocide as they also incorporate different modes of settlement, faith and culture'.[82]

Alastair McIntosh makes the point that 'a world replete with nuclear weapons could turn a struggle for diminishing resources into an altogether more deadly encounter involving whole national populations'.[83] This is one of the most credible scenarios for climate crisis-related genocide.

In summary, there is evidence that climate impacts will not automatically generate conflict, but the impact of sea level rise in particular is so far under-examined. Therefore, it is impossible to dismiss the argument that, combined with other pressures in weaker, less developed states, climate impacts could become important factors driving conflict.

Humanitarian assistance

The pre-established norm of humanitarian assistance (including intervention) is a third way in which climate crisis will help to reshape social and political relations. Flooding and extreme weather events are likely to create crisis situations demanding immediate, unconditional, politically neutral policy action to rescue endangered populations and prevent further loss of life. In the era of climate change, such activity is likely to be much more frequent and on a larger scale. A complex and in many ways highly effective system of humanitarian response exists under UN auspices with significant contributory assistance from INGOs and local NGOs. Such action, led and resourced by the global Western state, while far from sufficient, is increasingly sophisticated and organised to act immediately or even pre-emptively to prevent and limit suffering whether resulting from natural disasters such as earthquakes or tsumamis or manmade from wars or genocide, through a system of humanitarian response.

The three main climate-determined drivers explored here have obvious overlap and interconnection: human rights are intimately bound up with war and genocide, being the pre-eminent contexts for massive and systematic violation that so stirs human outrage and disgust. The relationship between humanitarianism and human rights and responses to both is a complex matter, complicated by the misuse of terminology, it is often difficult to disentangle two distinct phenomena, the humanitarian essentially being a subset of the broad concept of human rights. 'Humanitarian intervention', normally associated with a forcible breach of sovereignty by external actors, has been, in most historical cases, associated with human rights violations on a massive scale. The war in Kosovo (1998–1999) is one of the better-known examples. There was a need for immediate humanitarian action to protect and care for victims of mass forcible deportation, but the underlying problem was one of systematic human rights abuse of the majority Albanian population by the Serbian state over a period of years. Cyclone Nargis in Burma in 2007 resulted in the massive devastation of the Irrawaddy delta region and perhaps 150,000 lives were lost. The fact that the ruling junta refused, for a considerable period, much-needed international humanitarian assistance highlighted the longstanding structural impediment of systemic human rights violations in one of the worst rights-violating regimes.

State theory challenges us to see the state as a power centre considerably developed by traditional Weberian understandings. In *The Sources of Social Power*, Michael Mann has emphasised the way in which states, through polymorphous crystallisation, form in different ways at distinct moments and in relation to various issue areas. I have argued elsewhere that the Western state is regularly re-cast through such a process.[84] In Bosnia in 1992, initially the Western state crystallised into an entity dominated by British and French doubts about involvement, Simms' 'pusilanimous realism', with an isolationist bent in the Bush administration content with European leadership on a European crisis.[85] Only by 1995 did the Western state re-crystallise into a more US-driven form able to respond effectively to the Srebrenica and other massacres of July/August 1995. During the 2011 Libya crisis the global Western state crystallised in a unique manner, with many roles reversed. It seems the Arab uprisings, particularly in Libya, have themselves contributed to the development of the global Western state – through its re-crystallisation during the actual NATO intervention (France and Britain led the US and then a small array of EU powers to mobilise the 'global layer of state' and NATO), through the (partial, ongoing, yet highly fraught) democratisation of Libya and through perpetuation of that revolution, for example, witnessed in Libya's material support for revolution in Syria.[86] How might such polymorphous crystallisation work in climate crises?

Climate crises will shape and be shaped incrementally by the emerging global state but it may only be an unprecedented humanitarian crisis that will presage significant new structural change.[87] Typically political leaders tend to avoid dealing with the signs of an impending major crisis until such an event actually happens. As Mastrandrea and Schneider note, the IPCC did not assign 'subjective probabilities to climate warming either within or outside the 1.5 to 4.5°C' predicted range.[88] Consequently, political leaders can argue that impacts are imprecise, too uncertain to expend huge resources in attempting to tackle either the root causes or take adaptive measures. But difficulty lies in attributing probabilities relating to 'every aspect of projections of climatic changes' that need to be recognised. There are uncertain probabilities in relation to 'the cascade of emissions, carbon cycle response, climate response, and impacts'.[89] And there remain, for example, 'uncertainties surrounding climate sensitivity'[90] and future population and prediction levels of economic development (and associated technology). For Helm, even the possibility of a catastrophe – a low probability, high-impact event – will be subject to rational analysis.[91] The cost of such an event can be calculated: the probability of the event – a small number – multiplied by the cost of resulting damage – a large figure. Given that such a catastrophic event could be many decades, even centuries away, applying the discount rate to the resulting figure, suggests action is unlikely. As McKibben has argued, citing police brutality against civil rights activists, the emotional impact of climate change needs to be felt, 'if we are to take the giant steps we must take soon'.[92] Probably any affecting impact needs to be much greater than even that suggested here; otherwise, the efforts are likely to be piecemeal. The emergent global state is unlikely to respond adequately until a major event occurs that dwarfs, for example, the 2004 tsunami.

One such event might be a refugee or refugee-driven conflict crisis on an unparalleled scale. Biermann and Boas argue for 'global adaptation governance' to tackle protection and resettlement issues in such crises that could, as in Bangladesh, lead to conflict and genocide.[93] Myers estimates that the total number of people at risk from sea-level rise by 2050 is likely to be 162 million. In addition, 50 million people could become refugees due to droughts and other climate change impacts. What does it mean for the global Western state that more than 90% of the projected 302 million people being flooded by storm surges by the 2080s will be located in Africa or Asia?[94] Water scarcity could cause up to 1.8 billion people there to suffer from water deprivation, with 200 million experiencing climate change-related hunger by the 2050s. Returning to massively overpopulated Bangladesh, we find a country with the eighth largest population in the world, though with only the geographic space equal to the size of Nicaragua; their population is predicted to stabilise at 340 million. Given the low-lying situation of much of its land mass, its mega-delta flooding crises and cyclone zone status, it is very likely that rising sea level combined with river flooding (from glacier melt) and cyclones of greater intensity will produce massive refugee flows that, if not 'governed' carefully, will result in conflict either internally, as in the CHT, or with surrounding 'receiving' states, India and Burma.[95]

Conclusion

In an era of rapid, arguably revolutionary change, global state theory provides a more evolved reading of the current condition of state/governance development, engaging fully with the changes we are living through, widely recognised as 'globalisation'. The old cliché of the weakened nation-state by-product is recognised as one, rather small element of the substantial change that has happened and is continuing to drive state development. It explains 'the West' – a term frequently taken for granted and misunderstood in part, by governance theorists, the media, the public, and even Western political leaders themselves – as a much more complex political and social structure, constantly evolving and being re-cast as political and social change occur. What is most impressive is the explanation of the dynamic factors – involving global social relations based in significant part on human rights, whether in war or other contexts – that drive this development of politics and society. This article argues that climate crisis will generate similar impacts and transformative pressures.

There remains considerable uncertainty regarding many of the material consequences of increasingly certain climate change, let alone what social effects they might produce. But given the mutually constitutive nature of the developing sense of human society on a world scale, the global democratic revolution and the emergent global state, as social effects including war and genocide start to become apparent (and obviously climate-determined), humanity has a largely unrecognised hybrid state-form able to address issues of this magnitude in a manner which will focus on human rights issues central to climate impacts.

Notes

1. The emergent 'Global State' of Shaw's theory, is built on this Western conglomerate or network of state-based power, which it is argued, is increasingly global in its reach. A more complete elaboration of this idea and key underpinning terms and concepts will evolve below.
2. Gabriela Kutting, *Global Environmental Politics: Concepts, Theories and Case Studies* (London: Routledge, 2011), 194.
3. See for example, James Hansen, 'Obama's Second Chance on the Predominant Moral Issue of this Century' *Huffington Post*, 5 April 2010. http://www.huffingtonpost.com/dr-james-hansen/obamas-second-chance-on-c_b_525567.html.
4. Amy Sinden, 'Climate Change and Human Rights', *Journal of Land, Resources and Environmental Law* 27, no. 2 (2007): 255–71.
5. Dimitris Stevis and Valerie Assetto, 'History and Purpose in the International Political Economy of the Environment', in *The International Political Economy of the Environment* (Boulder, CO: Lynne Rienner, 2001): 239–56.
6. John von Neumann, quoted in John Vogler and Mark Imber, *The Environment and International Relations* (London: Routledge, 1996), 4–5.
7. John Vogler, 'The Environment in International Relations: Legacies and Contentions', in *The Environment and International Relations*, ed. John Vogler and Mark Imber (London: Routledge, 1996), 6.

8. Given existing particular preoccupations of leaders and publics, underpinned by strong national, religious and ethnic identities which consistently hamper the relatively minimal efforts of the UN system to ensure cooperation on policy areas critical to humanity, such as R2P, Security Council reform or even climate change itself, the notion of a legitimate, democratic top-down government on a global level does indeed seem some way off.
9. Martin Shaw, *Theory of the Global State*, (Cambridge: Cambridge University Press, 2000), 230.
10. Readers will note that a significant limitation to Shaw's theory is the emphasis on the political at the expense of the economic. The interconnection between global economic forces in the form of transnational corporations and the elites that direct them on the one hand and the political leaders of state and supra-state organisations on the other is not explored. Clearly, this is important to any discussion of climate change, driven as it has been, and continues to be, by the forces of global liberalisation of trade and markets that comprise the economic globalisation emphasised by traditional globalisation theory. There is not space to consider this aspect of development in this article.
11. Shaw, *Global State*, 26.
12. See John Barry and Kerri Woods, 'Environment', in *Human Rights: Policy and Practice*, ed. Michael Goodhardt (Oxford: Oxford University Press, 2009), 327, in which the authors make this point forcefully. See also Sinden, 'Climate Change', 259; Environmental Justice Foundation (EJF), *A Nation Under Threat – The Impacts of Climate Change on Human Rights and Forced Migration in Bangladesh* (London: Environmental Justice Foundation, 2012); Audrey Chapman, 'Integrating Human Rights and the Environment Within the United Nations' (2002). http://shr.aaas.org/hrenv/docs/wssd12_2001.htm; and Karim Ahmed, 'Environmental Protection, Public Health and Human Rights: An Integrated Assessment' (Global Children's Health and Environment Fund National Council for Science and the Environment, April 2003). http://shr.aaas.org/projects/human_rights/reports/Ahmed2003.pdf.
13. The Fifth Assessment Report will be published in 2014.
14. The impact of developed states is relatively minor and is, for reasons of space, not considered here. Various criticisms have been levelled at the work of working group two of the AR4 on climate change impacts. Most notable was Professor J. Graham Cogley, supported by Michael Zemp from the World Glacier Monitoring Service, who raised the highly damaging issue of the melting of the Himalayan glaciers, suggesting that the IPCC got the date (2035) wrong by 300 years. While the IPCC authors denied the claims, it has subsequently distanced itself from the claim. See Pallava Bagla, 'Himalayan Glaciers Melting Deadline "a Mistake"', 5 December 2009. http://news.bbc.co.uk/1/hi/world/south_asia/8387737.stm. There has also been criticism regarding the proportion of the Netherlands below sea level and African crop yield projections and concern about the use of 'grey literature' from non-peer reviewed sources.
15. IPCC, *Fourth Assessment Report 4 Climate Change (Synthesis) 2007*, 45. http://www.ipcc.ch/publications_and_data/ar4/syr/en/main.html.
16. IPCC, *Climate Change*, 45; see also Sheila Watt-Cloutier, 'Climate Change and Human Rights', *Human Rights Dialogue*, series 2, issue 11, Spring 2004.
17. IPCC, *Climate Change*, 46.
18. Ibid.
19. BBC News, 'Philippines Storm Kills Hundreds in Mindanao Floods', 17 December 2011. http://www.bbc.co.uk/news/world-asia-pacific-16229394.
20. IPCC, *Climate Change*, 48. There was very high confidence about this. It was also noted that coastal located industries would be vulnerable.
21. Ibid.
22. IPCC, *Climate Change*, 49.
23. Ibid.
24. Ibid.
25. Ibid.
26. Ibid., 50.
27. Ibid.
28. Ibid.
29. Ibid.
30. Ibid., 52.
31. Ibid., 53.
32. Ibid., 53–54.

33. Ibid., 54.
34. Ibid.
35. Bill McKibben, *The End of Nature* (London: Viking, 1990). In Kevin Anderson and Alice Bows in 'Beyond "Dangerous' Climate Change: Emission Scenarios for a New World', *Philosophical Transactions of the Royal Society*, 29 November 2010, 41, the authors go further, arguing that there is now 'little to no chance of maintaining the rise in global mean surface temperature at below 2°C, despite repeated high-level statements to the contrary'. Impacts associated with 2°C have been revised such that it 'more appropriately represents the threshold between dangerous and extremely dangerous climate change'.
36. Munich Re, for example, the largest reinsurer in the world, has publicly stated its expectation for the near future of an exponential increase in natural disaster-related costs, according to Stephen Schneider and Janica Lane, 'An Overview of "Dangerous" Climate Change' (2005): 1. http://stephenschneider.stanford.edu/Publications/PDF_Papers/Schneider-lane.pdf.
37. See Naomi Oreskes, 'Beyond the Ivory Tower: The Scientific Consensus on Climate Change', *Science* 306 (5702): 1686, in which the author has analysed the scientific literature on climate change. Oreskes has analysed 928 abstracts of papers from refereed scientific journals between 1993 and 2003, concluding that there is a scientific consensus on the existence of anthropogenic climate change. Seventy-five percent of the abstracts either explicitly or implicitly accepted the consensus view; the remaining 25% dealt with methods or paleoclimate, thus taking no position on current anthropogenic climate change. None of the abstracts disagreed with the consensus position. There are a few dissenting voices on climate change. Richard Lindzen of MIT and Freeman Dyson of the Institute for Advanced Study question the accuracy of IPCC projections. Others, including Tim Patterson of Carleton University and David Douglass of the University of Rochester, argue that climate change has been caused by natural processes. Robert C. Balling of Arizona State University and Judith Curry of Georgia Institute of Technology argue the cause is not yet established. A limited number of other dissenters, including Craig Idso and Patrick Michaels of Arizona State and the University of Virginia respectively, argue effects are more limited than the IPCC proposes.
38. Schneider and Lane, 'An Overview of "Dangerous" Climate Change', 1.
39. Shaw, *Global State*, 18.
40. Ibid., 17.
41. Ibid., 161.
42. Ibid., 16.
43. Ibid., 161.
44. D. Potter, D. Goldblatt, and M.Kiloh (eds.) *Democratization* (Cambridge: Polity, 1996): 7 cited in ibid.
45. Shaw, *Global State*, 194.
46. Ibid., 167.
47. Ibid., 168. There are four main ways in which these events are revolutionary: they are fundamental challenges to structures of authoritarian states; they are rooted in widespread popular agency; they are connected to changes in national structures and international relations; and these changes 'show every sign of durable impact on the forms and character of power'. They are not primarily socio-economic revolutions, but represent historic political transformations. Global-democratic revolution is not simply about the form of government within any nation-state, but rather the shape of world order, ibid., 169.
48. Shaw, *Global State*, 199–200.
49. Ibid., 213.
50. Ibid., 215.
51. Ibid.
52. Nils Petter Gleditsch, 'Whither the Weather? Climate Change and Conflict', *Journal of Peace Research 49* 1, no. 3 (2012): 3–9. See also Halvard Buhaug, 'Climate Not to Blame for African Civil Wars', *Proceedings of the National Academy of Science* 107, no. 38 (2010): 16477–82.
53. Joshua Goldstein, *Winning the War on War: The Decline of Armed Conflict Worldwide* (New York: Dutton, 2011).
54. For example, Steven Pinker, *The Better Angels of our Nature* (New York: Viking, 2011).
55. See also Andrew Mack, 'Global Political Violence: Explaining the Post-Cold War Decline', *Human Security Report Project Policy Papers*, March 14, 2007. http://www.hsrgroup.org/

IsakSvensson; Andrew Mack, 'From Bombs to Banners? The Decline of Wars and the Rise of Unarmed Uprisings', *East Asia Security Dialogue* 42, no. 3 (2011): 219–37.
56. The forthcoming fifth assessment report working group on impacts (due after March 2014) is expected to address these issues in greater detail.
57. Including Idean Salehyan, 'From Climate Change to Conflict? No Consensus Yet', *Journal of Peace Research* 45, no. 3 (2008): 315–26. Wario Adano, Ton Dietz, Karen Witsenburg and Fred Zaal, 'Climate Change, Violent Conflict and Local Institutions in Kenya's Drylands', *Journal of Peace Research* 49, no. 1 (2012): 65–80.
58. Thomas Bernauer, Tobias Bohmelt and Vally Koubi, 'Environmental Changes and Violent Conflict', *Environmental Research Letters* 7 (in press), quoted in Gleditsch, 'Whither the Weather?', 3.
59. Halvard Buhaug, Nils Petter Gleditsch and Ole Magnus Theisen, 'Climate Change and Armed Conflict' (paper for the Department of Energy/Environmental Protection Agency workshop on Research on Climate Change Impacts and Associated Economic Damages, Washington, DC, 27–28 January 2011).
60. Ragnhild Nordas and Nils Petter Gleditsch, 'Climate Change and Conflict', *Political Geography* 26, no. 6 (2007): 627–38.
61. David Zhang, C.Y. Jim, George C.-S. Lin, Yuan-Qing He, James J. Wang and Harry F. Lee, 'Climatic Change, Wars and Dynastic Cycles in China Over the Last Millennium', *Climatic Change* 76, no. 3–4 (2006): 459–77.
62. Richard Tol and Sebastian Wagner, 'Climate Change and Violent Conflict in Europe Over the Last Millennium', *Climatic Change* 99, no. 1–2 (2010): 65–79.
63. Gleditsch, 'Whither the Weather?', 3.
64. Adano et al., 'Climate Change, Violent Conflict and Local Institutions in Kenya's Drylands', 77.
65. Cullen Hendrix and Idean Salehyan, 'Climate Change, Rainfall, and Social Conflict in Africa', *Journal of Peace Research* 49, no. 1 (2012): 35–50.
66. Tor Benjaminsen, Koffi Alinon, Halvard Buhaug and Jill Tove Buseth, 'Does Climate Change Drive Land-Use Conflicts in the Sahel?', *Journal of Peace Research* 49, no. 1 (2012): 108.
67. Thomas Bernauer and Tobias Siegfried, 'Climate Change and International Water Conflict in Central Asia', *Journal of Peace Research* 49, no. 1 (2012): 227–40.
68. Jaroslav Tir and Douglas M. Stinnett, 'Weathering Climate Change: Can Institutions Mitigate International Water Conflict?', *Journal of Peace Research* 49, no. 1 (2012): 211–25.
69. Conor Devitt and Richard SJ Tol, 'Civil War, Climate Change and Development: A Scenario Study for Sub-Saharan Africa', *Journal of Peace Research* 49, no. 1 (2012): 129–45. This sub-Saharan Africa scenario study found that the effects of both civil war and climate change on economic growth have been underestimated.
70. Rune Slettebak, 'Don't Blame the Weather! Climate-Related Natural Disasters and Civil Conflict', *Journal of Peace Research* 49, no. 1 (2012): 163–76.
71. Clionadh Raleigh and Dominic Kniveton, 'Come Rain or Shine: An Analysis of Conflict and Climate Variability in East Africa', *Journal of Peace Research* 49, no. 1 (2012): 51–64.
72. Jeffrey Sachs, 'Crisis in the Drylands', *Scientific American*, 17 January 2008. http://www.scientificamerican.com/article.cfm?id=crisis-in-the-drylands. See also, Valerie Percival and Thomas Homer-Dixon, 'Environmental Scarcity and Violent Conflict: The Case of Rwanda', *Occasional Paper, Project on Environment, Population and Security*, (Washington, DC: American Association for the Advancement of Science and the University of Toronto, June 1995). http://www.library.utoronto.ca/pcs/eps/rwanda/rwanda1.htm.
73. Dan Smith and Janani Vivekananda, 'A Climate of Conflict. The Links Between Climate Change, Peace and War', *International Alert*, November 2007, 12.
74. Smith and Vivekananda, 'A Climate of Conflict', 7.
75. Ibid., 3–4. See also Jürgen Scheffran, Tobias Ide and Janpeter Schilling, 'Violent Climate or Climate of Violence? Concepts and Relations with Focus on Kenya and Sudan', *The International Journal of Human Rights* 18, no. 3 (2014): 369–90.
76. Mark Levene, 'Predicting Genocide in an Age of Anthropogenic Climate Change: An Interim Report', 2009. http://www.crisis-forum.org.uk/.
77. Ibid.
78. CNA, *National Security and the Threat of Climate Change* (VA: CNA Corporation, 2007), 5. http://www.securityandclimate.cna.org/.
79. International Institute for Strategic Studies, *Strategic Survey* (London: International Institute for Strategic Studies, 2007), 24.

80. Smith and Vivekananda, 'A Climate of Conflict', 8.
81. Ibid.
82. Jonathan Ward, 'Climate Change and Genocide: Will Anthropogenic Climate Change Increase the Risk of Genocide in Africa?' January 2008. http://www.crisis-forum.org.uk/
83. Alastair McIntosh, 'Where Now "Hell and High Water"?', *ECOS – Journal of the British Association of Nature Conservationists* 30, nos 3–4, December 2009. http://www.crisis-forum.org.uk/.
84. Gregory Kent, 'Defining Europe, Defining the West: The Balkans and Crystallisations of Power', Crucible Centre for Human Rights Working Paper, April 2011.
85. Brendan Simms, *Unfinest Hour. Britain and the Destruction of Bosnia* (London: Penguin, 2002).
86. The case of Syria appeared to be replicating this development of British and French initiative ahead of US involvement. However, at the time of writing, the British parliament's rejection of military action, followed by President Obama's focus on the elimination of Syria's chemical weapons capability – a not to be dismissed achievement however – has threatened arguably to undermine any crystallisation of Western power in an important case of asystematic human rights violation by a state on a massive scale.
87. This assumption – of the continued existence and development of the global state – is based on the historical argument that war and massive human rights violation are the key impetus for the cooperation that underpins the development of the global state. Although war can be said to be in decline, it is far from eradicated. Massive human rights violation is far from a purely historical phenomenon. Climate change is another factor that ensures cooperation into the future.
88. Stephen Schneider and Michael Mastrandrea, 'Probabilistic Assessment of "Dangerous" Climate Change and Emissions Pathways', *Proceedings of the National Academy of Science* 102, no. 44 (November 1, 2005). http://www.pnas.org_cgi_doi_10.1073_pnas.0506356102, 13–14.
89. Schneider and Lane, '"Dangerous" Climate Change', 8.
90. Ibid., 8.
91. Dieter Helm, *Climate Change Policy* (Oxford: Oxford University Press, 2005), 5.
92. Op.-Ed., 'Indifferent to Planet Pain', *New York Times*, 4 September 1999; See also Ronnie Lipschutz, *Global Environmental Politics, Power, Perspectives and Practice* (Washington, DC: CQ Press, 2004).
93. Frank Biermann and Ingrid Boas, 'Preparing for a Warmer World: Towards a Global Governance System to Protect Climate Refugees', *Global Environmental Politics* 10, no. 1 (2010), 60.
94. Ibid., 69.
95. Permanent displacement of 70 million Bangladeshis is a commonly used figure.

Culturecide in changing Micronesian climates? About the unintenionality of climate change

Rebecca Hofmann

Rachel Carson Center for Environment and Society, Ludwig-Maximilians University, Munich, Germany

This article establishes and explains the (cultural) genocidal dimension of climate change, engaging aspects of identity, intellectual property and cultural heritage from an emic view of the people concerned, namely the inhabitants of Chuuk, Federated States of Micronesia. The Pacific Islands are widely recognised as being specifically vulnerable to the impact of a warming climate. The threat it poses to the scarcely, but partly densely populated land resources might climax in the need to evacuate the islands altogether in the not so distant future. As identity in these communities is intrinsically linked to locality, the erosion of land also threatens to destroy people's cultural roots. Although a society's cultural self-conception is covered in a whole collection of declarations and conventions, the connection between climate change and its impacts, including migration dynamics and (cultural) genocide, has so far largely been ignored. Here, cultural change and migration are a dynamic part of Micronesian people's daily life. However, the consequences of current environmental changes, alongside the prospect of permanently losing the as yet well-guarded links to their homes, are likely to set in motion a re-evaluation of cultural self – a process that deserves global attention and legal consideration.

The Pacific Islands have become a symbol of the pending catastrophe of increasing sea levels due to global warming. The majority of the Pacific Islands are tiny in size and hardly rise more than two or three metres above sea level. The high volcanic islands are generally steep with only a small coastal rim on which most of the construction and population are concentrated. Such topographies are easy prey for erosion, resource degradation and destruction of the built environment caused by torrential tropical rains, tidal waves and storm surges, which are likely to increase in frequency and intensity with climate change.[1] Pictures of nature's destructive power over the population's food sources and acquainted environments, driving local people out of their homes without the prospect of return, now figure prominently in the media.[2] Yet, environments are not only a physical entity, but are perceived by humans and modulated into cultural values, norms and other aspects that, in short, constitute a particular culture.

'Cultures live and cultures die. Cultures live by the transmission of law, knowledge, land, and resources from one generation to the next', as legal scholar Danielle Conway[3] states in an article on customary law and identity. The image of a dead culture is extreme and not convincing as postcolonial, diaspora and globalisation discourses on 'invented traditions'[4] or 'traditional futures'[5] show. Yet, the question of what happens if tradition in its literal meaning – transmission – is interrupted on account of the loss of one of the aforementioned components, the lack of recipients, or the generational link where transmission takes place being cut, is one that deserves some attention in light of climate change. Global warming has the potential to induce all of these threats; hence, to investigate genocide in relation to this phenomenon is not as implausible as it might seem at first. The term '(cultural) genocide' was coined by the Polish-Jewish law professor Raphael Lemkin, author of the draft convention on genocide, who dedicated his life to lobbying for the inclusion of cultural genocide into the Convention on the Prevention and Punishment of the Crime of Genocide, established in 1948.[6] To Lemkin, 'culture is as vital to group life as individual physical well-being'[7] and genocide thus does not require physical violence to classify it as such. 'Genos', as Lemkin envisioned it, has to be seen as a 'collective' object[8] with physical and socio-cultural properties. Nevertheless, cultural genocide was omitted from the final convention as being too vague a concept, running the risk of lessening the atrocity of physical or biological genocide.[9] Also problematic are attempts to draw a line between cultural genocide and the societal dynamics of cultural change allowing many indigenous societies to successfully survive colonialism and other outside influences. Yet, if indigenous peoples perceive agents of change, such as human-induced environmental transformations, as a threat to their collective well-being, such changes should be regarded as a disturbance of procedural dynamics[10] and thus a breach of rights. In many indigenous societies, and certainly in Micronesia, people define themselves according to the relations they have with kin and land. Therefore, to evoke and consider the role of culture and environment in identification processes, the term 'relationality' seems fitting.[11] The destruction of relationality leads then to what Claudia Card calls 'social death'.[12] Social death alone can cause severe psychological distress without harm to the physical body of a society. She further correlates social death to cultural death and comes to this conclusion: 'If my hypothesis is correct, the term "cultural genocide" is probably both redundant and misleading – redundant, if the social death present in all genocide implies cultural death as well, and misleading, if "cultural genocide" suggests that some genocides do not include cultural death.'[13] The intentionality she addresses is a question that has also caused Damien Short to see land claim issues in Australia as a 'series of continuing genocides'.[14] Where the formation of relationships within a collectivity and their cultural values are disrupted, the group's 'distinct existence' is endangered, as Short[15] describes. While he concludes that such social death could be the unintentional by-product of political, economic or missionary expansionism, the forcible diminutions of a people's sovereignty in any form are not.[16]

This article does not in any way attempt to equal human-induced climate change with actions directed towards the disruption of a people's relationality and less even with the inconceivability of a deliberate intent to extinguish 'others'. It explores, instead, possible links between climate change and associated loss of 'cultural identity' in Micronesia.[17] It proposes to re-conceptualise the physical environment as an inherent part of Pacific Islanders' identity and relationality, in spite of high rates of migration and the representation of Pacific-belonging as depending mainly on social parameters. As indigenous Pacific scholars stress, no matter how distant from the islands people travel, the bond is always upheld and carefully guarded.[18] This article will therefore look at global climate change as a new form of land alienation with the potential to negatively impact physical as well as cultural environments, using the example of Chuuk in the Federated States of Micronesia.[19] Land here is the guiding principle for social organisation and personal identity. In the words of *longue durée* inhabitant Father Hezel[20]: 'If land was life for traditional Micronesians, then loss of land was a form of death'; in Chuuk, according to one author[21] it was lamented with the cry 'I am no longer alive.'[22] Where land and 'social vitality'[23] are an entity, the erosional impact of climate change on nature might also lead to an erosion of culture. To explicitly distinguish between cultural genocide as a direct or indirect, yet active and deliberate, cultural destruction on the one side and climate change as a catastrophic environmental transformation without lethal intentionality on the other side, the term *culturecide* is introduced. Just as genocide can take place without physical mass murder, a culture can be severely disrupted by the destruction of single elements such as land. While this is still in accordance with cultural genocide,[24] the difference lies with intentionality and scale. This means, whether the dire physical consequences for people, as for example in land development projects, are fully known or not, makes a difference. Albeit the initial nescience of the fatal consequences of greenhouse gases have long been overcome by extensive research on climate change and its impacts, the scale on which the global society on the one hand and each of its individual members on the other hand is involved, through economies and lifestyles, requires a new dimension of thought. And it would certainly go too far to speak of a 'coordinated plan of different actions aiming at the destruction of essential foundations' of collective life.[25] Culturecide nevertheless also alludes to the question of climate change justice and points towards the discourse of many (indigenous) groups who see ongoing air pollution and environmental deterioration clearly as human rights violations. Here, culturecide specifically relates to the congruent destruction of 'habitat', an acknowledged dimension in climate change rhetoric and partly conceptualised as ecocide. Apart from a strictly eco-systemic meaning, however, a habitat is also the space in which the lives of distinct cultural groups take place. While human life in its entirety is addressed in ecocide discourses, culturecide focuses more directly on human relations to and understandings of the environment and is therefore preferred for the purposes of this article. Through climate change, spaces and inhabitants face new and challenging implications, especially in regard to cultural patterns of explanation and reaction, and thus of continuity. Moreover, the neologism of culturecide might also be a way around the ongoing debate on the interdependency of cultural destruction and physical violence in cultural genocide.[26] Therefore, the main argument for using culturecide instead of cultural genocide or ecocide is the spatial, temporal and intentional disconnection of the emission of greenhouse gases and its impact, which nevertheless requires due recognition as a threat to the holistic existence of many human societies.[27]

Cultural identity and its protection: a brief overview

Marginalised and indigenous people worldwide fear losing control over their culture and traditions as symbols of identification[28] and relationality.[29] 'Culture', hereby, has become an overly used term and more often than not its definition is left unclear or vague. In brief, culture can be understood as a strategy used by groups to organise and understand their life and environs. Together with material objects, shared meanings, norms and values build a frame of reference for identity, shaped by generationally accumulated and transmitted experiences and practices of a social collective. Cultural traditions and related knowledge should be recognised and respected as a binding element between a group's past, present and future, representing a group's heritage.[30] It should further be acknowledged that such knowledge has owners or is held in trust by particular individuals or collectives.

Attempts to grasp the core of 'culture' and to provide a legal standardising body that protects every aspect of it are as manifold as the definitions and terms themselves. The quest to define 'what is cultural heritage and how to best protect it' was thus a long and complicated one, using catchwords such as natural heritage, folklore, living human treasures and masterpieces.[31] In October 2003, these ideas were eventually brought together in the adoption of the Convention on the Safeguarding of Intangible Cultural Heritage, defining intangible cultural heritage as 'the practices, representations, expressions, knowledge, skills – as well as the instruments, objects, artefacts and cultural spaces associated therewith – that communities, groups and, in some cases, individuals recognize as part of their cultural heritage'. Two important points need further examination: The concept of cultural space was included 'to mean a place in which popular and traditional cultural activities are concentrated, but also a time generally characterized by a certain periodicity (cyclical, seasonal, calendar, etc.) or by an event'.[32] Heritage 'is constantly recreated by communities and groups in response to their environment, their interaction with nature and their history, and provides them with a sense of identity and continuity' (Art. 2, sec. 1). Thus, place and nature are explicitly seen to be part of the formation of identity and the constitution of group relationships. Maybe a bit surprising at first, but even the UN Convention on Biological Diversity (CBD) recognises this link and stresses the intrinsic cultural value of biological diversity, affirmed in reversal by UNESCO's 2001 Universal Declaration on Cultural Diversity.[33] Thus, unlike in Western binary ideas of culture versus nature, in most indigenous ontologies the two spheres are not separated. Land, environment and culture are an interwoven and interdependent whole, building a system of cultural ecology in which human beings interact with their surroundings and in which their concepts of the world have evolved and to which they relate their place in the universe.[34] In the Pacific, the loss of traditional ethno-biological knowledge was addressed by the government of the Federated States of Micronesia in its National Biodiversity Strategy and Action Plan of 2002 and the theme with which the 2010 International Year of Biodiversity was celebrated – picturesquely boiled down to *Value Island Biodiversity, It's Our Life*. Life encompasses identity and thus culture. In Pacific Island societies, both are strongly connected to land as the grounding of genealogies. Alienation of territory should then be given special attention and must be understood as much more than economic deprivation.[35] Instead, alienation means estrangement from strings of collective identity in which individual personalities are embedded. As Micronesian scholar Joakim Peter explains in reference to his Western Pacific island in Chuuk State, 'We need to belong to places, the physical plots, taro fields, coconut groves, sandy beaches, portions of reefs, fishing corals, and the island in general'.[36] What frightens Micronesians, according to Peter, is to be *paselo* (or *pasónó*)[37] – to be adrift in unfamiliar waters, which cannot be connected with memory, where people do not have control of location, direction or movement. The navigational metaphor also illustrates Micronesian time conception resembling a loop more than the Western linear understanding. In order to steer the canoe to a travel destination, the sailors look back and align the canoe to what is behind. In the diaspora, people also look back and emotionally indulge in memories of moonlit beaches, embody place through food sent from home, and keep close ties to kin and clan via modern telecommunication systems while they steer through unfamiliar waters.

In the end, the document that comes closest to a round and sound idea of an indigenous cultural identity and that gives out a comprehensive list of what to protect is the UN Declaration on the Rights of Indigenous Peoples of 2007. It guarantees the human right to cultural existence (summed up in art. 31, sec. 1) and honours the contribution of indigenous cultures to the common heritage of humankind. Namely the importance of land is again and again repeated and stressed implicitly as much as explicitly (i.e. articles 25, 26 and 27).

It clearly states that people have the right to *their own lands, territories, and resources* (art. 8, sec. 2(b)), to *sites* (art. 12, sec. 1), and to the *transmission* of all that comprises tangible and intangible property and heritage; thus they are linked to 'past, present and future manifestations of their cultures' (art. 11, sec. 1). Moreover, the language of the declaration challenges all member states to actively engage in the protection and conservation of the lands and environments of indigenous communities and asks for just and fair compensation in cases of any land alienation or forced relocation.[38] Interestingly, while the initial 1994 draft included the term cultural genocide, the final version includes the idea of genocide (art. 7, sec. 2), but its cultural aspect has been reduced to prohibiting the destruction of culture (art. 8, sec. 1). Nonetheless, the document's biggest flaw is that it is merely a declaration and, as such, impeachments remain without consequences. Hence, the rights to land and resources and traditional knowledge connected to them as a crucial part of cultural integrity are recognised as a human right and are affirmed in international and national policies.[39] Unfortunately, however, many of those statements are mere declarations or, when they have legal value, are signed by states without any recognition of individual or criminal responsibilities – traits that much resemble climate change legislation.

Chuuk, a perishing paradise?

Chuuk, in the Caroline Islands of the Western Pacific, is an almost-atoll with several high islands (*chuuk* meaning mountain in the local language) in one of the world's largest ocean-lakes, fringed by an outer barrier reef of some 200 km in length. Outside the lagoon, several low atolls form the *fánáápi*, the sandy lands of Chuuk-State. The list of physical dangers and threats is long, and various waves of history have seen inter-island warfare, four different colonial powers, evangelisation and globalisation. No matter how turbulent the waters of time have been, these islands have remained a home to people. In the words of Micronesian political scholar Carl Heine: 'In time of war, and in time of peace, this was their home. Though foreign flags were raised, and treaties and trust agreements signed, this was their home.'[40] And now, climate models predict that these precious bits of land may become uninhabitable within the next 50 years – less than a lifetime. It is thus hardly surprising that the island states started to lobby the Western world to protect their homelands as early as 1992 at the UN meeting in Rio.[41]

However, vulnerabilities lie not on the physical side alone; they are also 'rooted in a people's understanding of the ontology and cosmology of indigenously perceived climate processes and how transformations in these ontological and cosmological notions threaten their cultural order'.[42] It is this cultural order or identity under the threat of climate change that has to be analysed and for which we need to 'translate the scientific data we have into descriptions of what it means for society'.[43] The pillars of Micronesian identity ascription are the relations to kin and land as a man from Satawan, an outer island of Chuuk, explains:

> We ground our identity in the land of our ancestors and the stories and genealogies connected with it. On my island, there is some *faaw*, you know, some block of coral. It's always been there and it is said to belong to this man called Aaw. You can even see his footprint on it and his grave is a small islet a little ways from our island. There is a story to it, of course, and we all know about it. To us, it's like one of your monuments or statues. (Personal conversation, January 2012)

The personal self is strongly linked to a person's collective identity, which again is largely determined by kinship, traditionally embracing the extended family and lineage (matrilineal descent group) as well as clan membership, securing ties beyond flesh and blood. Although this relationship is slowly breaking down,[44] relationships are still more vivid than what most Westerners experience in their own families. The bond is illustrated and repetitively renewed through the prominent aspect of sharing (*néwúnéwúffengen*, 'to own or use collectively' or simply *ach*, meaning an inclusive 'our(s)'), which touches upon every facet of life. An important feature is the sharing of songs and stories that ensures the continuation and affirmation of collective identity. Stories about how clans came into existence and how they were established on the land they now live on, plant a person's roots and give a feeling of belonging. The landing of people in certain environs thus constitutes the very beginning of historical awareness. Thus, it is not so much territorial boundaries that are seen and guarded as intellectual property, but the identification process through ancestral genealogies. And though Micronesian communities have high emigration rates, it is essential to the community that part of the family stays back to take care of the land. This sense of responsibility for and everlasting belonging to their land is carried with them wherever they go and is most evident in death. Bodies are transferred home, regardless of cost, to lie in lineage land, symbolically claiming title to it for the deceased's offspring.

Land is also indirectly an attribute of Chuukese identification as illustrated by a 29-year-old who, when asked what made him Chuukese, answered that it was 'the language and our culture' (personal conversation, January 2012). To a subsequent question about what makes up their culture, he said: 'The way we eat. We eat big portions. When we have food, we eat a lot of breadfruit, taro, banana, fish.' How are language and food linked to land and identity? The language as identity marker is well recognised and *fóósun Chuuk* has a much more detailed way of expressing kinship relations as well as an extensive food vocabulary. Beyond that, *fóósun fénúwan*, which translates into the language of the land, as part of an exclusive version of the language formerly used by chiefs and traditional priests contains a whole list of food and other nature metaphors. Similarly, food also serves as an important symbol of identity, which in Chuuk coheres to principles of 'you are what you eat', that is local food that comes from the land, and with whom you share your food. Thus, both language and culture, expressed in the way they talk and eat, are elements of Chuukese personhood and they stand in direct relation to land. This identification with home and land through food remains regardless of distance, as demonstrated by the tons of pounded taro, breadfruit and salted fish that are sent to relatives craving the taste of home but living in neighbouring Guam and also to Hawaii and the US mainland.

Thus, the Chuukese socio-cultural universe, their awareness of time and history, are encoded and contextualised in multiple threads, all of which are anchored in a sense of place. Localities have continuing value due to their bridging function from history to present day. Sites of importance are often also associated with power invested through the presence of ancestor spirits or other ghosts. The tales and meanings connected with such sites continue to be cornerstones of the people's identity even within today's Christian society, and as such, the reading of this cultural landscape could almost be seen as the Micronesian version of our visit to a *Heimatmuseum* (museum of local history). To sum up, land, by all means and in all its ecological and cultural meanings, needs to be protected from destruction – in this case, from the impact of climate change.

Culturecide in Chuuk?

The consequences of climate change are representing a slow, yet irreversible threat to the islands' physical and, thus, cultural existence. The energy of waves, tides, storm surges, rains and winds are natural erosive factors on tropical islands; the increase of their dynamics, however, is owed to anthropogenic global warming and changes to their regimes are observed by local populations. During the months of November 2011–January 2012, three exceptionally high tides were recorded by the inhabitants of Chukiyénú on Toleisom (better known by its abbreviated name of Tol) and of other villages on other islands in Chuuk. Whereas the body of water is pushed towards the shores with greater might during the windy season, known by the name of *kúún ennefen* (translated as the trade or north wind season, or the time of little breadfruit and high waters), 'this tide now, we would not call *kúún ennefen*. This tide is even higher, so we are scared, we don't know why it happens. Now, it's coming over the road, into our houses. The high tides come more often now' (Chukiyénú elder, November 2011).[45] The main road on Toleisom, leading around the island, is a path of coral rocks and dirt and had to be closed several times during these high tides. Boat piers were inundated as much as the houses that are situated too close to the shore. Several people specifically lamented the flooding of their *fanang*, the cooking houses (small, tin-roofed constructions lacking walls with stamped ground on family compounds), where they store firewood and prepare food. But the cooking houses are also important socio-cultural places where people meet and talk. While some men pound taro or breadfruit on the ground, others sit around and have conversations, exchange the latest news or gossip, waiting their turn; the same can be said for the women and their tasks. Even when the fires are cold, this is a popular place for all generations to come together. It seems a substitute, at least to a certain extent, for the traditional *wúút* (men's or meeting houses) in the men's case and the streams where women used to wash the clothes, before 'washing machines put an end to the laundry gossip' as one Chuukese humorously commented (personal conversation, January 2012).[46] With water encroaching coastal family compounds, it touches yet another element of great importance in the Chuukese culture – the graves of the families' ancestors. They are customarily buried on lineage land[47] and are, in the case of emigrants, brought home with a lot of effort, sometimes depleting a family's savings to their financial ruin. While the increasing erosion of land washes away traditional food (taro swamps and coconut and breadfruit trees), landmarks and clan ground – symbols of Chuukese cultural identity – the question remains what will happen if people actually have to leave their islands for good. While research on Micronesian transnationality thus far has been centred on motives such as health care, education, job opportunities and lifestyle allurements, debate is now shifting to questions of involuntary or unwanted migration due to climate change. Speaking in terms of culturecide, one of the central questions has to be if cultural spaces will be relocated along with the people or if people will be left uprooted without the soil that has been nurturing their socio-cultural self over generations. While resettlement itself does not necessarily lead to the extinction of a culture, it tends to restructure and accommodate cultural values appropriating them to the new locale, carrying the potential to leave people not just *dis*placed but, more importantly, very much *mis*placed. Oliver-Smith explains:

> Loss of livelihood and the capacity to sustain oneself endanger individual and social identity, producing a loss of status and resulting in marginalisation and social disarticulation through the fragmenting of social networks such as kin, religious, and other collaborative groups and associations. The loss of a house is also the loss of a social 'place'. The entire community or social world is thus endangered by such individual losses.[48]

Moreover, climate change-induced movements also play into heritage discourses as driving 'increasing numbers of people away from their home places, forcing them into an uncertain journey with little more than a suitcase filled with the artifacts of a stranded past', just as Apfelbaum[49] observes in the case of political and economic migrations. It is this 'movement away from the known past, through an altered present, toward an uncertain future'[50] that has the power to rob people of the trust in their collective cultural self. Whatever they have learned and experienced, consciously and unconsciously, loses its anchor, and upon touching new and unknown ground will have to find a new harbour. Scholarly debates on 'climate induced migration', however, largely evolve around Western security and legal questions or adaptation issues that ignore the complex reality of migration.[51] What it leaves out is any discussion in terms of legislation regarding the safeguarding and protection of people's culture, intellectual and intangible property, or heritage. In Micronesia itself, more effort is currently put into the retraction and conservation of American and Japanese Pacific history (namely in the form of memorabilia and contemporary witnesses' testimonies of the Pacific war) than into any consideration of the people's destiny should they lose the spatial reference to their cultural practices of remembrance.

Although, or because, culture is a critical element in a society's abstraction of identity and *Weltanschauung* (world-view), it is often overlooked or taken for granted without sufficient differentiation. Francis Hezel, long-time director of *Micronesian Seminar*, a non-governmental organisation (NGO) that has been engaged in public education for over 40 years and has established a database on Micronesian islands, brings in another interesting and critical examination of mostly Western approaches to culture. In *Loss of Culture*, he admonishes his readers to not forget where culture belongs, that is 'within the people who pass it along – and who change it, in big ways and small, as they are forever doing'.[52] It is a warning to check the true moral motives behind scientific debate, lobbying and policy making that revolve around the question of whom cultural safeguards really serve, and who profits from collecting and enlisting the world's cultural diversity as part of a common human heritage. The very concept seems to be born from a Western idea of having to preserve the past, separating and storing it away from the very actions that once shaped it. Yet, the notion of safeguarding knowledge is quite strong in indigenous societies where protection mechanisms play into the politics and power of knowing. A large part of cultural knowledge, especially when it touches the spheres of specialists, is secret or only accessible to certain people who prove themselves worthy. In Chuuk, the older generation may lament the loss of culture, but many of the traditional knowledge holders explicitly decide to take their mental possessions to their grave. In Chuuk, traditional knowledge is known as *rong* and in order to keep *Rong from Wrong*, to cite the title of an insightful paper by Nason and Peter (2009),[53] elders intentionally interrupt transmission due to the fact that, in their view, the next generation is either unworthy or simply uninterested in learning the old ways. For much of the younger generation who grew up with store-food, trained physicians and technical devices, oral knowledge has become obsolete, reduced at best to being nice entertainment or valued as something that is customarily part of daily life but on which they no longer depend. Additionally, cultural modification and alteration is as old as culture itself and it can be argued that it is this flexibility that makes cultures so persistent through centuries of colonialism and globalisation. Thus, dynamic reactions between transformation and continuity build the backbone of cultural resilience, instead of fixing culture in a static picture that has to be preserved. Henry et al. conclude:

> In fact, the concept of heritage requires a notion of 'change'. Heritage is a construct that serves a human (not necessarily universal) impulse to 'fix' or hold on to something that is in reality continuously changing, whether due to natural forces such as the weather or to anthropogenic

> forces (including war, modernization and economic development and so on). Thus, cultural practices and ideas, objects, places, land and seascapes become valued as heritage particularly in the face of their potential loss.[54]

Obviously, cultural heritage gains importance when placed at risk. This also explains why certain cultural practices are sometimes adhered to more vigorously in the diasporic experience. To inherit something necessarily includes the act of passing it on, which, in Micronesia, can take half a lifetime. All in all, heritage and its transmission carries a strong element of social and cultural binding, that is, between the master and the apprentice or from a mother transferring her land (and all connected genealogies and perceptions of the living and spiritual space) to her daughters, reinforcing the family's legacy. Places of importance are thereby nested in cultural spaces and are not simply loci to commemorate the past. Instead, they draw importance just as much from recent powers and related beliefs or from common-day practices. In this regard, Petersen offers an interesting reading of Nan Madol, the 'mysterious city of stone'[55] in Pohnpei, the municipal island state of the Federated States of Micronesia. He suggests that there is much more behind the stone structures than mere mythology; they stand as an ongoing reminder of the evil and corruption of the centralised power of Nan Madol's mystical founders. In this way, it teaches Pohnpeians moral and political values on an everyday basis. Yet, as a 'contested landscape', this demonstrates differing interpretations of cultural heritage. To the Pohnpeians, trimming the vegetation that threatens to swallow the ruins and other Western conservation measures equals an insult to the structures' power of resistance that it has shown in enduring centuries of tropical humidity and storms. Likewise, King, after working as a consultant for the Micronesian Heritage Program for two years, came to following insight:

> Preserving historic places, cultural places, cannot be isolated from the rest of culture. The only way preserving historic places made sense to traditional Micronesian people was as part of an overall program of cultural preservation that valued bread-fruit and coconut and pandanus harvesting and processing, cooking in earth ovens, fishing with spears at fishtraps, gathering shellfish. The program had to respect traditional spiritual beliefs and esoteric knowledge (...). It had to respect rights to land, trees, and the reef.[56]

The concept of culture and heritage is thus dynamic and holistic and Hezel[57] has a good point in arguing that the 'Pacific Islands are not a living museum for the entertainment and edification of outsiders, to remind them that the world still contains uncomplicated places with warm, friendly populations; they are the home of thousands of people who must bravely face the future, just as the people of other nations must.' The advocates of cultural genocide, including Lemkin, warn not to confuse cultural change and cultural genocide,[58] so what has climate change to do with the destruction of culture and why does it demand attention as proposed in this article?

Conclusion

In accordance with the anthropological understanding of 'relationships and social networks' as the 'basis of personal and social identity',[59] the concept of social death[60] views the loss of social vitality as the loss of identity, diminishing the meaning of one's existence. It redirects the focus from body counts to the loss of relationships, from body to soul, that represents the shift from genocide to cultural genocide. In his analysis on Micronesian personality in relation to adoption, Rauchholz[61] stresses the importance of belonging, thus of social vitality, in order to maintain an uncontested personhood. While one could argue that locale, then, is less important, locality and genealogy are interdependent variables in Micronesia as illustrated by clan founding legends. Hence, while cultural genocide is understood as the systemic extermination of cultural values as framed by Lemkin, culturecide, on the other hand, is distinguished on the grounds that environmental change due to global warming is neither an active nor intentional, nor planned systemic destruction of culture. It rather has to be regarded as a negative 'side effect' of a process that, although set in motion by human action, was never intended or understood to have such destructive powers.

It is widely acknowledged that Small Island States and developing countries are subject to special conditions, asking for both extra attention and support in the face of climate change. This special need is often associated with the uncertainty of potentially climate-driven physical impacts as well as with the limited scope of adaptation options. A population's disposition and capability to adapt much depends on local beliefs and values.[62] With the increasing impact of climate change, these energies are likely to reach a threshold beyond which culturally transmitted strategies are no longer sufficient. While research of climate change impacts is almost inflationary and whereas scholars of culture and law eagerly engage in intangible or intellectual property rights, a connection between these rights and climate change is largely undiscussed. Therefore, the identification of critical thresholds of climate change and its physical impacts should be supplemented by studies of socio-cultural thresholds of identity and adaptation capacity.

In sum, this article outlined the importance of the intrinsic link between land and identity in Chuuk, a discussion that can be transferred to other Micronesian communities and beyond. It further enumerated laws recognising this link that is immediately threatened by climate change impact. Due to its unintended and indirect nature, a concept such as cultural genocide will hardly be aligned to the consequences of global climate change. Francioni, however, points out that the 'destruction or the desecration of symbolic objects and sites that are essential to the enactment of a people's culture (. . .) is a violation of their collective dignity no less than a violation of their personal dignity'.[63] And as such, the cultural degradation and loss of relationality experienced by peoples at risk certainly touches on human rights. The new climate regime has the potential to disconnect the spatial relation of people and their homelands – to cut the spatial connection of a person's self-conception which is embedded in a socio-cultural collective. To ignore these links by evoking today's widespread transnationalism of Pacific Islanders means to undermine the symbolic valuation with which people still invest their identities. And while mobility resembles a well-tested and proven adaptation strategy and traditional resilience pattern, the shift from voluntary and short-term to involuntary and final puts both the physical and the psychological resilience at stake, especially when 'deprivation and displacement shake the foundations of personal worldview and identity, [and] challenge the culturally constructed vision where the world is a place imbued with logic and where life makes sense'.[64] The question thus remains whether or not climate-induced displaced islanders or 'climate refugees'[65] will become cultural orphans or if they will undergo a cultural metamorphosis. This once again leads to the question of how far cultural genocide, or here culturecide, is a mere construction of scientific deliberation. Micronesians may be innovative enough to make up for the loss of their islands by incorporating even such dramatic changes into their flowing system of conception, but land ties may also be too strong a condition and their loss might lead to irreversible disruptions of the socio-cultural psyche. So far, the reassessment of the self in the diaspora seems to trigger a renaissance of collective cultural identities and their heritage enacted not least through the perpetuation of active ties to the islands.

Thus, while the West should be careful not to lament the possible loss of these paradises because it projects its own desires on them, the prospect of eternally losing the place of one's childhood memories and collective belonging does trigger universal human feelings of devastation. In the end, climate change not only furthers the erosion and destruction of physical land, but also of human cultural diversity, albeit unintentionally.

Acknowledgement

This work was supported by the German Federal Ministry of Education and Research [01UV1010A-B].

Notes

1. For a detailed and recent report on the US affiliated islands see PIRCA, *Climate Change and Pacific Islands* (ANU Newsroom, 2008).
2. See, for example, '"Climate Change Has Sealed our Fate" in Kiribati'. http://news.anu.edu.au/2008/06/25/climate-change-has-sealed-our-fate-kiribati-president/ (accessed 17 April 2013); or Cooney Campbell, 'Bougainville Works to Relocate Climate Refugees' (Australia: ABC News Online). http://www.abc.net.au/news/2009-02-13/bougainville-works-to-relocate-climate-refugees/295182 (accessed 12 June 2013).
3. Danielle Conway, 'Indigenizing Intellectual Property Law: Customary Law, Legal Pluralism, and the Protection of Indigenous Peoples' Rights, Identity, and Resources', *Texas Wesleyan Law Review* 15 (2009): 207.
4. Eric Hobsbawm and Terrence Ranger, *The Invention of Tradition* (Cambridge, MA: Cambridge University Press, 1983).
5. James Clifford, 'Traditional Futures', in *Questions of Tradition*, ed. Mark Salber Phillips and Gordon Joel Schochet (Toronto: University of Toronto Press, 2004), 152–68.
6. For more information on Raphael Lemkin, see, for example, Dominik J. Schaller and Jürgen Zimmerer, 'Raphael Lemkin "On Genocides"', in their guest edited issue of the *Journal of Genocide Research* 7, no. 4 (2005).
7. Damien Short, 'Australia: A Continuing Genocide?' *Journal of Genocide Research* 12, no. 1 (2010): 47.
8. Ibid., 50.
9. It is interesting to note the simultaneous development of the Universal Declaration of Human Rights, particularly the overlap of eliminating cultural genocide from the former and minority rights from the latter (see Johannes Morsink, 'Cultural Genocide, the Universal Declaration, and Minority Rights', *Human Rights Quarterly* 21 (1999): 1022), leaving behind a gap in the protection of minorities and cultural identity, nominally covered by arguments that other future and existing regulations will serve better to the task.
10. Short, 'Australia: A Continuing Genocide?', 52.
11. See for example, Sabine Hess, *Person and Place. Ideas, Ideals and the Practice of Sociality on Vanua Lava, Vanuatu* (New York; Oxford: Berghahn Books, 2009), for a good account of Pacific place and relationality.
12. Claudia Card, 'Genocide and Social Death', *Hypathia* 18, no. 1 (2003): 63–79.
13. Ibid., 63.

14. Short, 'Australia: A Continuing Genocide?', 62.
15. Ibid., 52.
16. Ibid., 62.
17. The islands that are home to approximately 110,000 inhabitants are spread over seven million square kilometres with large bodies of water in between. Yet, the water might be separating the islands and inhabitants as much as it is connecting them; Micronesia is thus a recognised cultural area amongst anthropologists and other scientists (Manuel Rauchholz, 'Towards an Understanding of Adoption, Person, and Emotion. The Ideal Norm and Reality of Life Amongst the Chuukese in Micronesia' (unpublished PhD diss., University of Heidelberg, 2009), 150).
18. See works by Joakim Peter for Chuuk, Sa'iliemanuLilomaiava-Doktor for Samoans and Epeli-Hau'ofa for emic views on spatiality. Much has also been written by Western scholars; see, for example, the monograph series edited by Jürg Wassmann and Katharina Stockhaus-White, *Person, Space and Memory in the Contemporary Pacific* (New York; Oxford: Berghahn, 2007).
19. In 2011–2012, the author spent 11 months in Chuuk, one of the four states of the Federated States of Micronesia, teaching at a local college and doing anthropological research (including interviews, informal conversations and participant observation) on questions regarding human–nature relationships in changing climates. All data and quotes, when not indicated otherwise, stem from the author's fieldwork and are her sole responsibility.
20. Francis Hezel, *The New Shape of Old Island Cultures – A Half Century of Social Change in Micronesia* (Honolulu: University of Hawaii Press, 2001), 34–5.
21. Patricia Parker, *Land Tenure in Trukese Society, 1850–1980.* (PhD diss., University of Pennsylvania, 1985).
22. For examples of forced relocation and the high psychological stress involved, for example in the case of Marshall Islanders who had to leave their home islands due to US nuclear testing, see Michael Lieber, ed., *Exiles and Migrants in Oceania* (Honolulu: University of Hawaii Press, 1977).
23. Card, 'Genocide and Social Death', 63. See also Conway, 'Indigenizing Intellectual Property Law', 207–56.
24. See Damien Short, 'Cultural Genocide and Indigenous Peoples: A Sociological Approach', *The International Journal of Human Rights* 14, no. 6 (2010): 831–46.
25. Lemkin cited in Short, 'Cultural Genocide and Indigenous Peoples', 836.
26. See, for example, the argument between Short and Dirk Moses in Short, 'Australia: A Continuing Genocide?'
27. For a discussion on the relation between climate change and ecocide, see the contribution by Martin Crook and Damien Short in this issue. Martin Crook and Damien Short, 'Marx, Lemkin and the Genocide–Ecocide Nexus', *International Journal of Human Rights* 18, no. 3 (2014): 298–319.
28. Toon van Meijl, 'Pacific Discourses About Cultural Heritage and Its Protection: An Introduction', *International Journal of Cultural* Property 16 (2009): 222.
29. This threat became reality for so-called 'lost generations', who underwent intensive Western education and alteration and lost all connection to their native culture, i.e. Australia's 'stolen generation' or Micronesia's youth growing up in America's Navy and later Trust Territory regimes where store-food became abundant and formal school education compulsory, rendering the students without any knowledge of traditional food preparations and little material culture.
30. Van Meijl, 'Pacific Discourses', 224.
31. UNESCO Convention Concerning the Protection of the World Cultural and Natural Heritage 1972; UNESCO and WIPO 1982: Model Provisions for National Laws on the Protection of Expressions of Folklore against Illicit and Other Prejudicial Actions; UNESCO 1989: Recommendation on the Safeguarding of Traditional Culture and Folklore; Living Human Treasures 1993; Proclamation of Masterpieces of Oral and Intangible Heritage (UNESCO 1997, which, for example, listed the Vanuatu Sand Drawings that are by their proper nature easy prey for higher sea levels).
32. 155 EX/Decision 3.5.5. Annex Regulation (c) as cited in: Noriko Aikawa, 'An Historical Overview of the Preparation of the UNESCO International Convention for the Safeguarding of the Intangible Cultural Heritage', *Museum International* 56, no. 12 (2004): 141.
33. Art. 1 proclaims that 'as a source of exchange, innovation and creativity, cultural diversity is as necessary for humankind as biodiversity is for nature' (as cited in: Francesco Francioni,

'Beyond State Sovereignty: The Protection of Cultural Heritage as a Shared Interest of Humanity', *Michigan Journal of International Law* 25 (2004): 18).

34. See for example, Philippe Descola, *Jenseits von Natur und Kultur* (Berlin: Suhrkamp, 2011).
35. Also see Michael Goldsmith, 'Who Owns Native Nature? Discourses of Rights to Land, Culture, and Knowledge in New Zealand', *International Journal of Cultural* Property 16(2009): 325–39, who states that land and culture for the New Zealand Māori are inalienable, or Marilyn Strathern, 'The Patent and the Malanggan', *Theory, Culture and Society* 18, no. 4, (2001): 1–26, who argues for including land in intellectual property in Melanesia.
36. Joakim Peter, 'Uleletiw: Imaging of My Paradise', in *Pacific Places, Pacific Histories. Essays in Honor of Robert C. Kiste*, ed. Brij Lal (Honolulu: University of Hawaii Press, 2004), 261.
37. Ibid., 273.
38. To the author's best knowledge, this point has so far not been analysed in regard to its possible applicability to the otherwise broadly conducted debates on climate change-related movement, whose actors are usually labelled as 'climate refugees'.
39. UN Committee on Economic, Social and Cultural Rights; Committee on the Elimination of all Forms of Racial Discrimination; Declaration on Cultural Diversity.
40. Carl Heine, *Micronesia at the Crossroads – A Reappraisal of the Micronesian Political Dilemma* (Honolulu: University Press of Hawaii, 1974), 3.
41. John Connell, 'Losing Ground? Tuvalu, the Greenhouse Effect and the Garbage Can', *Asia Pacific Viewpoint* 44, no. 2 (2003): 90. In 1989, the First South Pacific Regional Environment Programme Meeting on Climate Change and Sea Level Rise took place in the Marshall Islands. One year later, after the World Climate Conference, the Association of Small Island States (AOSIS) was formed. However, as predicted catastrophes had not yet taken place, AOSIS lost interest in its lobbying business in the mid-90s (Connell, 'Losing Ground?', 90). The failure to see the connection between current environmental changes and global warming continues to create uncertainty and hinders the initiation of preventive measures.
42. Jerry Jacka, 'Global Averages, Local Extremes: The Subtleties and Complexities of Climate Change in Papua New Guinea', in *Anthropology and Climate Change. From Encounters to Actions*, ed. Susan A. Crate and Mark Nuttall (Walnut Creek: Left Coast Press, 2009), 207.
43. Markus Vogt, 'Climate Justice from a Christian Point of View: Challenges for a New Definition of Wealth' (talk, 'Religion in Global Environmental and Climate Change: Sufferings, Values, Lifestyles Workshop', Potsdam, 11–13 January 2010).
44. See, for example, Hezel, *The New Shape of Old Island Cultures.*
45. As later confirmed to the author, pronounced La Niña effects caused these events. If the El Niño–Southern Oscillation (ENSO) is intensified by climate change as predicted, people will have to deal with floods and droughts much more often in the years to come.
46. However, most households do not dispose of washing machines as only the main island, Wéénéé, is electrified, while few villages have solar power and most run diesel generators during evening hours.
47. Before Christianity, sea burials were also common practice as well as keeping body parts of the deceased within or underneath the house.
48. Anthony Oliver-Smith, 'Climate Change and Population Displacement: Disasters and Diasporas in the Twenty-First Century', in *Anthropology and Climate Change. From Encounters to Actions*, ed. Susan A. Crate and Mark Nuttall (Walnut Creek: Left Coast Press, 2009), 123.
49. Erika R. Apfelbaum, 'And Now What, After Such Tribulations? Memory and Dislocation in the Era of Uprooting', *American Psychologist* 55, no. 9 (2000): 1009.
50. Carla Roncoli, Todd Crane and Ben Orlove, 'Fielding Climate Change in Cultural Anthropology', in *Anthropology and Climate Change. From Encounters to Actions*, ed. Susan A. Crate and Mark Nuttall (Walnut Creek: Left Coast Press, 2009), 87.
51. The literature on 'climate induced migration' has grown considerably throughout the last couple of years and so has the number of scholars who look at the topic from various angles. Western discourses are nevertheless still prevailing. Here are some few examples: Frank Biermann and Ingrid Boas, 'Preparing for a Warmer World', *Global Environmental Politics* 10, no. 1 (2010): 60–88; Maxine Burkett, 'In Search of Refuge: Pacific Islands, Climate-Induced Migration, and the Legal Frontier', *Asia Pacific Issues*, no. 98 (2011): 1–8; as well as Betsy Hartmann, 'Rethinking Climate Refugees and Climate Conflict: Rhetoric, Reality and the Politics of Policy Discourse', *Journal of International Development* 22, no. 2 (2010): 233–46.

52. Francis Hezel, 'Loss of Culture. How Real Is the Threat?', *Micronesian Counselor* 56 (2005): 9.
53. James Nason and Joakim Peter, 'Keeping Rong from Wrong: The Identification and Protection of Traditional Intellectual Property in Chuuk, Federated States of Micronesia', *International Journal of Cultural Property*, 16, Special Issue Number 3 (2009): 273–90.
54. Rosita Henry and William Jeffery, 'Waterworld: The Heritage Dimensions of "Climate Change" in the Pacific', *Historic Environment* 21, no. 1 (2008): 12–18.
55. Glenn Petersen, 'Nan Madol's Contested Landscape: Topography and Tradition in the Eastern Caroline Islands', *Isla – A Journal of Micronesian Studies* 3, no. 1 (1995): 106.
56. Thomas King, 'How Micronesia Changed the U.S. Historic Preservation Program and the Importance of Keeping It From Changing Back', *Micronesian Journal of the Humanities and Social Sciences* 5, nos 1–2 (2006): 511.
57. Hezel, 'Loss of Culture', 2.
58. John Docker, 'Raphael Lemkin's History of Genocide and Colonialism' (paper for the United States Holocaust Memorial Museum, Center for Advanced Holocaust Studies, Washington, DC, 2004), 14–15. Using the example of American Indians, he sees cultural change as gradual adaptation to outside influences while cultural genocide refers to the complete and irreversible change, thus the destruction of a culture.
59. Oliver-Smith, 'Climate Change and Population Displacement', 123.
60. Card, 'Genocide and Social Death', 63–79.
61. Rauchholz, 'Towards an Understanding of Adoption, Person, and Emotion'.
62. Roncoli, Crane and Orlove, 'Fielding Climate Change in Cultural Anthropology', 87.
63. Francioni, 'Beyond State Sovereignty', 4.
64. Oliver-Smith, 'Climate Change and Population Displacement', 123.
65. Two good introductory readings on the debate of environmental and climate refugees are: Richard Black, 'Environmental Refugees: Myth or Reality?', *New Issues in Refugee Research*, no. 34 (2001): 1–19; and Etienne Piguet, 'From "Primitive Migration" to "Climate Refugees": The Curious Fate of the Natural Environment in Migration Studies', *Annals of the Association of American Geographers* 103, no. 1 (2013): 148–62.

Genocidal risk and climate change: Africa in the twenty-first century

Andreas Exenberger[a] and Andreas Pondorfer[b]

[a]Department of Economic Theory, Policy and History, University of Innsbruck, Austria; [b]Institute for the World Economy, the Environment and Natural Resources, Funding: BIOACID, Kiel, Germany

Climate change is often related to various adverse effects, among those are endangering food security and raising the risk of conflict. Some scholars go as far as identifying climate change as the main driver of civilisational crisis. But empirical evidence is rather inconclusive so far, particularly about its relationship to violence, and even more so, genocide. In this article, we provide a literature review of studies explaining certain forms of violence and especially the connections between climate change and violence as well as an empirical study about the connections of climate variables (temperature and rainfall) and agricultural production in sub-Sahara Africa. Further, we also provide an assessment of institutional risk factors given the historical record of sub-Saharan African states with respect to genocide and projections about the future development of agricultural production for the first half of the twenty-first century to also assess environmental risk. In doing so, we are able to identify countries of joint risk and promising directions of further research. In this context, we chose sub-Saharan Africa as our focus area for various reasons: first, because of widespread poverty, institutional weakness and dependence on rain-fed agriculture, sub-Saharan Africa is the macro region where the effects of climate change will very likely play out most adversely; further, it also assembles a pronounced historical record of violent conflict, notably including genocidal episodes; finally, mutual enforcement and endogeneity issues are particularly viable there, not the least in the form of a large potential impact of technological and institutional improvements.

Introduction

In 2007, judiciously connected to the Stern report on the economics of climate change and the fourth assessment report of the International Panel on Climate Change (IPCC),[1] Jürgen Zimmerer stressed the issue of 'environmental genocide' in an editorial of the *Journal of Genocide Research* and called for a 'fresh and wider look at the causes of genocide'.[2] Because works about climate and genocide remain nevertheless scarce, debate is still necessary, given the potential large-scale effects of climate change on societies world-wide, especially on poor regions lacking adaptive capacities and characterised by low opportunity costs of violence. For example, the United Nations Development Program (UNDP) predicted '[m]ajor losses in agricultural production leading to increased malnutrition and reduced opportunities for poverty reduction. Overall, climate change will lower the incomes and reduce the opportunities of vulnerable populations ... Developing countries are likely to become more dependent on imports from the rich world, with their farmers losing market shares in agricultural trade.'[3] But overall, while there is no doubt among scientists that accelerated climate change (especially global warming) is taking place and wide compliance that there is a considerable anthropogenic component to it, scientifically sound agreement has not thus far been achieved either about the relationships between observed changes and more complex environmental processes (like the occurrence of storms, for example) or of any related social processes.

One of the most distressing social processes is genocide. We glean a basic understanding of this violent act from the UN Genocide Convention of 1948 (intended destruction, in whole or part, of a national, ethnical, racial or religious group) with two relevant extensions: first, we limit our interest to mass killings and neglect other forms of genocidal violence (like the destruction of culture or systematic discrimination); second, we also add political groups as potential targets of violence (a phenomenon usually referred to as 'politicide'). Thus, we end up with a practical understanding of 'genocide' as geno- and/or politicidal mass killings. Further, because of the crucial role of intentionality, we usually observe genocide as a policy choice, more viable under certain conditions (particularly polarising forces), sometimes even being 'disturbingly rational'.[4] Consequently, while ideology plays an important role and psychopathic perpetrators have their place in genocide studies, most of the time people solve conflicts without turning to genocide, mainly because it simply does not make sense as a promising strategy from the viewpoint of potential perpetrators. Thus, the influence of climate change altering the framework in which political strategies are placed becomes increasingly relevant. In this context, we understand climate change as anthropogenic, continuous and sustainable warming (rise in average surface temperature) with associated phenomena (sea level rise, more frequent and more severe droughts and other extreme weather events, generally enlarged weather variability, etc.), which is increasing the relevance of water availability and hence the role of rainfall, especially in rain-fed subsistence agriculture. While climate change may result in extreme weather events, it is itself a long-term trend originating in the industrial revolution and increasingly effective in the last half-century.

In theory, direct and indirect causal links between climate change and conflicts are manifold.[5] It is often claimed that a changing climate will induce armed conflicts connected to increasingly scarce resources and flows of refugees from desertification and sea level rise. It is likely however, that there is a relationship between food security and violence; however, such a relationship is likely to be complex, strongly related to distributional and marginalisation patterns and nonlinear. As Nils Petter Gleditsch puts it, 'the effect plays out in interaction with exogenous conflict-promoting factors'.[6] Thus, it is hardly a surprise that empirical research about the conflict–climate nexus is not able to provide clear evidence. Quite the contrary, results are either inconclusive (sometimes to the point of counter-evidence) or sceptical about connections (remaining statistically insignificant). Hence, there is indeed a need of further research of all kinds: cross-country, case studies and systemic analysis. Our largely empirical contribution brings two strands of empirical literature together: the study of mass violence and climate change and the study of the effects of climate change on agricultural production, and hence results in a comparative assessment of institutional and agricultural risks on a country basis. Although this work is essentially explorative and it is not possible to predict genocides from its results (also because of the large relevance of contingency in this context), this will allow us both to look at a certain kind of combined danger and, thus, to contribute to genocide prevention by identifying countries where potential scarcity-induced crisis (a climate-related decrease of agricultural production) correlates with institutional factors of genocidal violence (regime types and forces of path-dependence).

Overall, the relationship between food security and violence is complex, strongly related to distributional and marginalisation patterns, nonlinear, and self-enforcing. Further, while political and socio-economic factors are likely to be more relevant than climatic ones, there is especially a considerable and often overlooked combined risk. At the same time, projections are very sensitive to assumptions about technological development and to political and economic improvements. This article aims to contribute to all these issues. In this context, we choose sub-Sahara Africa (SSA) as our specific case for two obvious reasons: first, it is a macro region with a diverse history of mass violence; second, its agricultural potential will be most negatively affected by climate change. Although the factors of influence regarding the investigated relationships are complex and not at all African 'by nature' (but much more linked to specific patterns of colonial extraction, prominently land-use patterns, export-oriented production, and the discrimination of certain population groups), they are particularly related to a high degree of vulnerability, connected with poverty and bad governance which result in a lack of resilience at the local and national level.[7] Further, climate change-related losses are suggested to be especially substantial in the case of smallholder farmers, still dominating SSA agriculture, where both cash-crop and subsistence production will be affected, limiting the capacity to achieve and sustain food security by a loss of available revenue as well as a loss of food. Finally, expert assessments support special emphasis on Africa. The geno-/politicide database of the Political Instability Task Force (PITF), for example, contains 128 out of 273 country-years from Africa, constituting 15 postcolonial SSA episodes.[8] Further, the NGO Genocide Watch provides a more extended list with more than a third of all cases after 1945 being from Africa. This also holds for their most recent 'countries at risk' list, published in April 2012: three out of nine 'extermination' cases are from SSA (Congo Kinshasa, Somalia, Ethiopia), as well as six out of eleven 'preparation' cases (Nigeria, Chad, Equatorial Guinea, Congo Brazzaville, Central African Republic, Guinea Bissau).[9] In sum, if there is a problem, it is very likely that it is urgent and relevant to Africa and this contribution provides information about which cases deserve or even require a second, more detailed, look.

This analysis proceeds as follows: the next section provides a review of studies about mass violence and especially connections with climate change; in the following section, we assess genocidal risk as the correlation between agricultural risk (the projection of increased food insecurity based on a cross-country analysis of the influence factors of agricultural production in SSA) and institutional risk (vulnerabilities related to political, economic and conflict-related factors); finally, we draw some conclusions and provide an outlook with respect to further research, which is especially necessary in the form of comparable in-depth case studies (while this article is largely about quantitative cross-country evidence). These conclusions confirm that the understanding of the issue is still insufficient, but that it is very likely that vulnerability is self-enforcing, that the interplay of political, socio-economic and climate factors is especially relevant, and that results are very sensitive to assumptions about technological development. What is beyond the scope of this article, however, is a detailed assessment of the impact of colonial patterns, particularly of land use, although these issues are certainly relevant to explain differences in vulnerabilities to climate change dynamics.

The empirics of mass violence and climate change

Quantitative cross-country studies about mass violence

There is a large and ever-increasing body of empirical literature about factors influencing the vulnerability to or the outbreak, intensity and duration of mass violence (sometimes in a binary form on the basis of conflict-years, sometimes in the continuous form of probabilities). A growing cottage industry of data mining and processing about conflicts has developed over the previous decades. Works begin from rather different perspectives, such as the study of interstate or civil war, of geno- and politicides, or of what Rudolph Rummel termed 'democide'.[10] The factors usually stressed encompass the economy (greed versus grievance), politics (autocracy versus democracy), demography (young men), culture (ethnic, linguistic, or religious fragmentation), geography (insurgency promoting terrain), and ideology (communism or racism).[11] In a recent quantitative cross-country civil war study, Paul Collier, Anke Hoeffler and Paul Rohner[12] summarise and recall influencing factors usually found to carry explanatory power: besides poverty and primary commodity exports (where nonlinearity and endogeneity have to be considered), they particularly find a lack of peace (short periods of peace between episodes of war), social fractionalisation (actually linguistic and religious), and the absence of an implicit French security umbrella (not being a former colony of France) to be statistically significant factors. Besides identifying and quantifying these factors of influence, the general and empirically well-grounded conclusion from this work is that civil war dynamics usually are more driven – from the perspective of (potential) rebels and in reaction thus also the governments – by appropriating resources than by overcoming injustice.

Quantitative studies about genocide have learned from these kinds of exercises. In her path-breaking study about political upheaval escalating (or not) into geno-/politicide between 1955 and 2001, Barbara Harff departed from the challenge 'to identify general conditions under which governments, and rival authorities in internal wars, choose such a strategy'[13] and identified six risk factors, which in combination raise the probability of escalation from about 3% (in absence of all risks) to about 90% (in presence of all six).[14] Harff's risk factors include a history of prior genocide, an exclusionary ideology from the elite, an autocratic government, a ruling elite from an ethnic minority, low trade openness, and a high magnitude of political upheaval. There is also a swing effect, most dramatically observed in Rwanda and Burundi: 'Case studies suggest that the more intense the prior struggle for power and the greater the perceived threat the excluded group poses to the new regime, the more likely they are to become victims of geno-/politicide.'[15] Finally, active discrimination against ethnic minorities contributes importantly to an explanation of the emergence of ethnic conflicts but much less to an explanation of why these conflicts escalate to genocide. However, there are also missing suspects in Harff's narrative: particularly poverty, which is relevant for upheaval but seemingly not for its escalation; external political links, which seem to be dominated by economic considerations; analysis of any colonial heritage; and a discussion about the relevance of post-colonial extractive patterns.

William Easterly, Roberta Gatti and Sergio Kurlat analyse the influence of income and (Western style) democracy levels in 163 episodes of mass killings of unarmed civilians from 1820 to 1998.[16] Generally, the results show that the occurrence and especially the number of victims increase with a decrease in income; this occurs to an even greater degree in cases of (Western style) democracy; though, in their analysis, the relationship with democracy is subject to several important qualifications (about data quality, statistical significance and causal relationships). In their words: 'We find that episodes of mass killing are more likely at intermediate levels of income and less frequent at the highest levels of democracy.'[17] In an excellent review of the democide and geno/politicide literature by Frank Wayman and Atsushi Tago,[18] the focus is on the onset of mass killings and the role of military governments. These authors also provide a methodological discussion about the implications connected to the respective data sets and definitions. They also raise doubts about the relevance of political indicators, especially in the case of geno-politicides. In this context, case study evidence is more reassuring: a qualitative comparative study of Rwanda and Botswana, for example, clearly supports the relevance of political inclusiveness (a democratic mediation of the political process) with respect to genocide occurrence.[19] While it is obvious that the discovery of diamonds in Botswana has supported wealth creation in comparison to Rwanda, it is not obvious from the onset that mining diamonds supports wealth dispersion (indeed, resources of this kind had the opposite effect in many other countries); instead, it was the political institutions of Botswana that ensured wealth dispersion.

The climate–violence nexus

While the empirical literature about violence is able to identify some causal factors, the literature about the relationship between climate change and violence is less conclusive. Jon Barnett and W. Neil Adger summarise that 'it is important to stress that climate change will not undermine human security or increase the risk of violent conflict in isolation from other important social factors'.[20] A more recent and more detailed study explicitly shows that for civil wars in Africa, climate-related factors are at best marginal compared to socio-economic, political and geographic factors: 'there is no direct, short-term relationship between drought and civil war onset, even within contexts presumed most conducive to violence. At the same time, the analysis solidifies claims of recent scholarship on the importance of ethnically inclusive institutions for maintaining peace.'[21]

On the other hand, given the high vulnerability of African societies to the forces of climate change, especially with respect to agricultural production (due to limited capacity in general and the large relevance of rain-fed agriculture in particular), one may go as far as comparing the African present with the European past. In this context, some scholars have claimed that climate change, via its impact on agro-economic systems, was the main driver of social and political upheaval in Europe between 1500 and 1800, peaking in the 'crisis of the seventeenth century'.[22] This is in line with claims in the IPCC assessment report of 2007 and the Human Development Report of 2007–2008 regarding more recent connections. These reports suggest that a loss of agricultural potential of 16.6% is predicted for SSA until the year 2080, even if potential positive effects from carbon fertilisation are taken into account (these numbers are fully in line with our results). Further, regional impacts are diverse, although positive outcomes at the country level remain a rare exception in Africa (applying only to Egypt and possibly Kenya).[23]

Nonetheless, endogeneity remains an issue, and not just a technical one, because socio-economic development is strongly linked to adaptation capacity. While industrialisation contributes (or at least has contributed) to growth and democracy (at the local level) and to climate change (at the global level), it also increases the resources available and improves the capacity for adaptation to its adverse effects, although with considerable distributional consequences.[24] Consequently, no effect of climate change on economic growth is found in countries with an income above the global median, while the impact is large in countries below the median, thus strongly affecting most of SSA.[25] Further, democracies are better able to organise environmental protection.[26] Additionally, these dynamics give way to vicious or virtuous circles of mutual enforcement. Conor Devitt and Richard Tol conclude that 'the impact of civil war and climate change . . . are sufficiently strong to keep a number of countries in Africa in deep poverty with a high probability'.[27] Two particular channels of negative influence have been pointed out: a deterioration of public resources to support the population and a reduction of the opportunity costs of rebellion. Particularly, 'armed conflict is more likely to occur in states where existing institutions and mechanisms for conflict resolution cannot provide people with the assurance that climate-induced economic problems will be resolved without recourse to violence'.[28] However, empirically there is only weak support for a mediating effect of political institutions and no support for a direct climate–conflict nexus via economic growth. Relationships are obviously more complex. Christopher Butler and Scott Gates argue that there is indeed a positive relationship between resource abundance, not scarcity, and conflict, given that there is a 'property rights bias' (differences in the access to property rights); they find that this relationship is intensified if differences in resource endowments are large.[29] By applying a broader concept of conflict, Cullen Hendrix and Idean Salehyan show that large deviations in rainfall result in increased social conflict in Africa; violent conflict is especially correlated to wetter years while non-violent conflict is correlated to drier ones.[30] A particularly relevant factor is the marginal status of a group: 'Environmental issues can be catalysts to low-level conflict in marginalized communities, but the critical factor is the extent of political and economic marginalization. Small, politically insignificant ethnic groups experience most conflicts related to environmental pressures.'[31]

Overall, while there is a complex and rather indirect relationship between climate change and violence, further conditions have to be met to let a situation of stress escalate into large-scale killing and, even more so, genocide. What most studies have in common is that they stress historical, political and economic reasons for the occurrence of genocide: a history of violence, a rather exclusionary political process and rather poor and exclusionary economic performance. In this context, a recent work by Busby et al. seems to be particularly worth mentioning.[32] By using the method of geographically mapping different insecurity risk factors in Africa, mainly on a sub-national level, they are able to show the large differences of vulnerabilities in geographic space and also the clustering of their occurrence. They especially focus on a combination of climatic and socio-economic factors, including both the aggregate and the household levels. Aside from several methodological caveats they extensively discuss in their study, the regions they identify as particularly problematic largely overlap with the countries identified in the rest of this article.

Assessing genocidal risk

Agricultural production and climate change in sub-Sahara Africa

We now turn to the effects of climate change on agricultural production in Africa by estimating an agricultural production function including labour, livestock, fertiliser, capital, land and irrigation, as well as precipitation, temperature and the incidence of droughts as input factors.[33] For the sample period (1961–2009) we assemble 2190 observations for 46 SSA countries.[34] There is considerable variation in all variables and a large degree of heterogeneity across the countries, pointing to the relevance of idiosyncrasies supporting the use of a fixed effects model. However, there is also a clear temperature trend and the five-year average in the period 2005–2009 is 0.67° Celsius higher than the 30-year average in the period 1961–1990, while rainfall development, unsurprisingly much more volatile, is more U-shaped, with particularly bad years in 1983, 1992 and 1990.

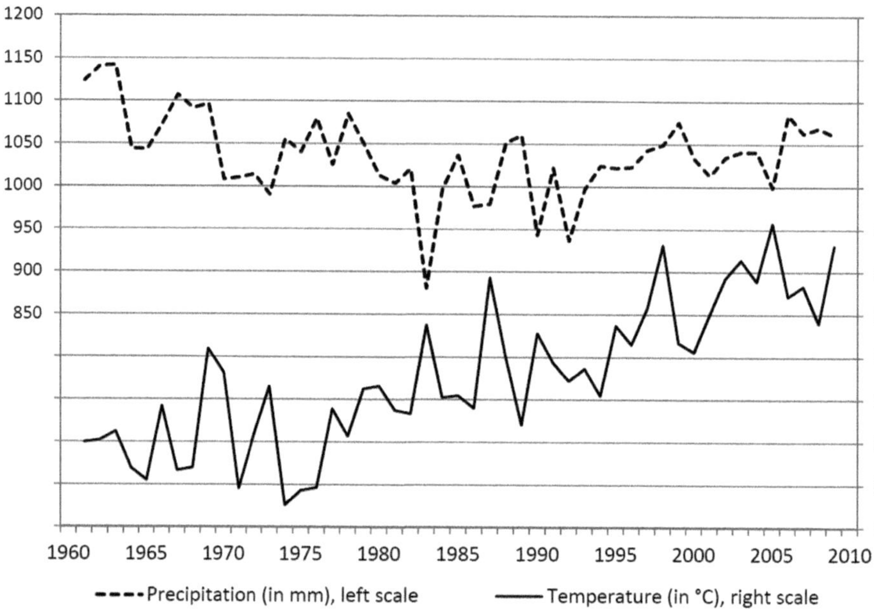

Figure 1. Temperature and precipitation trends in sub-Sahara Africa, 1961–2009
Source: See relevant text.

Table 1 presents the results of the multivariate standard regression, explaining the relevance of production factors influencing agricultural output. In model 1, a standard specification was estimated without climate variables and irrigated area.[35] As expected, all inputs have a positive sign (thus are stimulating output), and, except for modern capital (tractors), all inputs are statistically significant, with land being particularly relevant. Equally, the coefficient of traditional capital formation (livestock) is higher than the coefficients of modern inputs (tractors, fertilisers); this is linked specifically to investment and educational opportunities. Rainfall and temperature are also significant (although only weakly in the case of temperature). Further, in all models testing for rainfall (2, 3 and 5), rainfall emerges as a more relevant factor of production than the modern inputs, underlying the prevalence of rain-fed agriculture. Thus, decreases in rainfall experienced by large parts of SSA during the sample period reduced agricultural output while the effect of temperature was comparably marginal. There is also the expected negative effect of droughts on production, though this is not significant (despite model 4, when other climate variables are skipped). In model 5, irrigation is also included, but with weak results (to be explained by its little prevalence in sub-Sahara Africa).

Countries at risk, part I

The PITF database contains 15 postcolonial episodes of geno-/politicides in sub-Sahara Africa. There are many more contested cases, including most prominently killings in the context of the African World War, the communist regime in Ethiopia, continuing violence in Somalia and Ethiopia, the Eritrean-Ethiopian war, the Sierra Leonean civil war, and the Mozambiquean war of independence. To avoid discussion about the inclusion of contested cases, we stick to the restricted list of 15 episodes given in Table 2. While there is hardly any

Table 1. Agricultural production and climate change in sub-Sahara Africa

Dependent variable: agricultural production; time period: 1961–2009					
VARIABLES	(1)	(2)	(3)	(4)	(5)
Tractors	0.035	0.034	0.034	0.035	0.035
	(0.034)	(0.033)	(0.033)	(0.034)	(0.031)
Fertiliser	0.027**	0.029**	0.029**	0.027**	0.029**
	(0.012)	(0.012)	(0.012)	(0.012)	(0.012)
Livestock	0.274***	0.265***	0.265***	0.273***	0.265***
	(0.043)	(0.042)	(0.042)	(0.043)	(0.043)
Labour	0.298*	0.300**	0.302**	0.301*	0.262*
	(0.150)	(0.146)	(0.147)	(0.151)	(0.151)
Land	0.794***	0.788***	0.786***	0.792***	0.730***
	(0.220)	(0.214)	(0.214)	(0.219)	(0.226)
Irrigation					0.046
					(0.036)
Precipitation		0.159***	0.155***		0.162***
		(0.050)	(0.051)		(0.048)
Temperature		−1.334*	−1.322*		−1.222*
		(0.677)	(0.671)		(0.637)
Drought			−0.016	−0.027*	
			(0.015)	(0.015)	
Constant	−10.138***	−9.996***	−9.996***	−10.133***	−9.283***
	(1.818)	(1.740)	(1.739)	(1.813)	(1.844)
No. of observations[a]	1,894	1,894	1,894	1,894	1,894
R^2	0.823	0.829	0.829	0.823	0.831
No. of countries	46	46	46	46	46
Country fixed effects	Yes	Yes	Yes	Yes	Yes
Time fixed effects[b]	Yes	Yes	Yes	Yes	Yes
Cluster	country	country	country	country	country
F-test[c]	23.63	25.02	22.68	21.74	22.36

Notes: robust standard errors in parentheses; ***, ** and * constitute significance at the 1%, 5% and 10% levels, respectively.[a] 296 observations were dropped because of missing data about tractors and fertilisers.[b] Time effects are significant (Wald test).[c] The F-test refers to the input variables (time effects and input variables were tested separately).
Source: See relevant text.

doubt that the occurrence of mass violence is much more widespread in SSA than in only these cases, the problem is to identify *intended* and *discriminate* killing, which is indispensably necessary to make it genocidal.

Rwanda, Burundi, Angola, Uganda and Congo Kinshasa faced recurring genocidal violence.[36] The two largest cases presented are also connected to agriculture. In the case of the Rwandan genocide in 1994 (and also Burundi in 1993), there is hardly any doubt that the redistribution of land played a role in the killings, and thus, at least partially, also in their motivation.[37] But other factors clearly dominated and, aside from serious population pressure, there is not much sign of a relevant climate-induced deterioration of living conditions before the genocides.[38] Further, in the still heavily contested case of the Biafra war (1967–1970), starvation clearly was part of the military strategy of the Nigerian army to break secessionist resistance. Nevertheless, in both cases a factor related to climate became an instrument of, rather than a reason for, killing.[39]

Based on historical, economic and political reasons for mass violence, we assemble seven data series (form the years 1975–2009) in three categories to characterise SSA countries. The three categories are: (1) the historical record of violence (data on the

Table 2. Genocidal episodes in sub-Sahara Africa, 1960–2009

Episode	Duration (years)	Killings (total)	Brief description
Extreme cases (killings > 10,000 per month)			
Rwanda, 1994	0.3	750,000	Rwandan genocide
Nigeria, 1967–1970	2.7	1,225,000	Famishing strategy during Biafra war
Burundi 1993	0.3	48,000	Political struggle by massacre
Burundi 1988	0.1	12,000	Political struggle by massacre
Large cases (killings between 1000 and 10,000 per month)			
Uganda, 1980–1986	5.2	240,000	Anti-opposition government violence
Angola, 1998–2002	3.3	120,000	Civil war related genocidal campaigns
Angola, 1975–1994	19.1	546,000	Civil war related genocidal campaigns
Uganda, 1971–1979	8.3	216,000	Anti-opposition government violence
Rwanda, 1963–1964	0.6	12,150	Civil war related genocidal campaigns
Burundi, 1965–1973	8.3	129,750	Hutu-Tutsi political struggle
Smaller cases (killings < 1000 per month)			
Somalia 1988–1991	2.8	29,150	Anti-rebellion government violence
Congo Kinshasa, 1964–1965	1.0	6,150	Civil war related genocidal campaigns
Congo Kinshasa, 1977–1979	2.8	14,150	Anti-rebellion government violence
Ethiopia, 1976–1979	3.5	13,500	Anti-opposition government violence
Equatorial Guinea, 1969–1979	10.5	31,300	Anti-opposition government violence

Source: PITF (http://www.systemicpeace.org/inscr/PITFGenoPoliticides2012.xls); for number of killings: Anderton, 'Choosing Genocide', 464.

number of conflict years and average conflict intensities in conflict years); (2) the presence of economic difficulties (data on the number of years in recession, the average contraction in these years, and the 2009 log-level of gross domestic product (GDP) per capita in PPP$); (3) the presence of political suppression and exclusion (polity2 and polcomp scores, i.e. the degree of autocracy and the level of political competition). An overview of common characteristics of the countries concerned is given in Table 3. The table is already ordered by ascending levels of institutional risk: when all four of these risk factors are present, the country is ranked at the top; where none are present, the country is ranked at the bottom.[40]

Thus, we see that the countries regarded as particularly at risk share a combination of several risk factors: a history of recurring conflict, a large vulnerability to economic crisis, a lack of democracy, and a more exclusive political process, which all raise the risk of violence. And while Djibouti, Côte d'Ivoire, Mauritania and Congo Brazzaville are certainly debateable (as the 'violence factor' is hardly developed), the overall picture is clear. This is further supported when the occurrence of genocidal episodes (see Table 2) is surveyed: most of these occurrences happened in impoverished countries in some kind of political or economic crisis, while no events (excepting perhaps Uganda in 1980) happened in a democratic environment; further, none happened in times of economic prosperity and only three events occurred under competitive (but highly conflictual) political conditions.

Countries at risk, part II

For the second part of the risk section, we use our estimation for a projection of agricultural production until 2050.[41] We use the IPCC-SRES scenarios A1B and A2[42] and develop three projection scenarios (see Table 4).[43] Further, we simulate future climate change impacts by using the Hadley Centre Global Environmental Model applied to our data from the Tyndall Centre for Climate Change Research.[44] In the pessimistic scenario, A2,

Table 3. Countries at risk in sub-Sahara Africa

	(1)	(2)	(3)	(4)	(5)	(6)	(7)
Congo Kinshasa	25	3.6	231	20	-4.0	-5.1	2.4
Djibouti	4	1.0	2,061	21	-2.5	-4.4	3.3
Liberia	13	2.9	397	16	-4.2	-2.8	3.8
Rwanda	10	3.5	1,031	11	1.1	-5.8	1.6
Côte d'Ivoire	6	2.0	1,343	21	-0.4	-6.9	1.7
Mauritania	5	3.0	1,574	15	0.3	-5.9	1.8
Congo Brazzaville	6	2.0	2,220	13	1.1	-4.6	2.4
Uganda	30	2.9	1,152	12	1.2	-3.5	2.9
Burundi	15	3.7	368	20	0.2	-3.1	2.9
Sierra Leone	11	3.0	873	14	-0.3	-2.9	3.0
Ethiopia	21	5.3	684	15	1.2	-3.1	3.4
Nigeria	26	2.0	2,034	18	0.4	-0.9	3.4
Swaziland	0	–	3,440	13	1.3	-9.5	1.2
Eritrea	0	–	592	12	0.8	-6.5	2.0
Gabon	0	–	10,280	19	-1.1	-6.0	2.2
Tanzania	0	–	1,189	10	1.7	-3.8	2.5
Guinea	2	1.0	826	15	-0.1	-4.7	3.3
Cameroon	1	1.0	1,811	13	0.3	-5.9	3.6
Togo	0	–	734	18	-1.7	-4.7	3.6
Burkina Faso	0	–	902	14	1.0	-4.6	3.7
Zimbabwe	12	1.8	143	21	-2.9	-2.1	3.8
Angola	31	6.4	4,754	17	1.7	-4.5	3.8
Chad	24	3.8	1,276	16	0.9	-4.0	3.9
Guinea-Bissau	0	–	818	16	1.5	-2.1	3.7
Central African Republic	6	1.5	648	20	-1.2	-2.3	3.8
Niger	8	1.0	534	19	-0.4	-1.3	3.9
Kenya	6	1.5	1,206	15	0.4	-2.5	4.1
Mozambique	12	6.0	759	12	1.4	-1.7	4.7
Mali	6	1.0	999	11	2.1	0.3	4.9
Senegal	8	1.0	1,492	13	0.5	1.0	7.8
Lesotho	0	–	1,311	13	2.1	-0.3	3.9
Malawi	0	–	653	19	-0.8	-2.3	4.3
Zambia	0	–	1,765	22	-0.5	-1.7	4.4
Comoros	0	–	916	24	-1.1	0.3	4.8
Ghana	2	1.0	1,239	11	1.0	-0.3	5.1
Benin	0	–	1,116	14	0.5	0.4	5.5
Madagascar	0	–	753	20	-0.7	1.1	5.6
Cape Verde	0	–	3,779	6	2.9	3.2	5.8
Gambia	1	1.0	1,465	16	1.3	1.7	6.0
Botswana	0	–	8,868	11	4.2	7.0	9.0
Namibia	0	–	4,733	18	0.1	6.0	9.0
Mauritius	0	–	9,484	3	3.6	9.8	9.8

Notes and sources

(1) Conflict years: number of years in 'civil conflict' between 1975 and 2009; MEPV database (Monty G. Marshall, 'Major Episodes of Political Violence, 1946–2008', http://www.systemicpeace.org). (2) Conflict intensity: average intensity (from a low 1 to a high 10) in these years of 'civil conflict'; MEPV database. (3) Income: GDP per capita in 2009 in PPP$ (base year 2005, using chain rule); PWT database (Alan Heston, Robert Summers and Bettina Aten, *Penn World Table Version 7.0*, Center for International Comparisons of Production, Income and Prices at the University of Pennsylvania, May 2011. http://pwt.econ.upenn.edu/php_site/pwt_index.php). (4) Recession years: number of years with negative income growth between 1975 and 2009; PWT database. (5) Economic performance: average annual income growth (geometric mean) between 1975 and 2009; PWT database. (6) Democracy: average of the polity2 score between 1975 and 2009, i.e. the difference between the democracy and autocracy score (from an autocratic -10 to a democratic +10); Polity IV database (Monty G. Marshall and Keith Jaggers, *Polity IV Project: Political Regime Characteristics and Transitions*, 1800–2010. http://www.systemicpeace.org/polity/polity4.htm). (7) Political competition: average of the 'polcomp' score, between 1975 and 2009, which is denoting regulation and competitiveness of political participation (from a low 1 to a high 10); Polity IV database.

Table 4. Summary of the assumptions used for projections

	A2	*A1B*	*A1B_TT*
Tractor	fixed stock	fixed stock	fixed stock
Fertiliser	fixed stock	fixed stock	fixed stock + increasing elasticity (0.1 in 2050)
Area equipped for irrigation	none	none	fixed area + increasing elasticity (0.1 in 2050)
Livestock	livestock increases by 1.5 (base 2005–2009)	livestock increases by 2.5 (base 2005–2009)	livestock increases by 2.5 (base 2005–2009)
Agricultural area[a]	fixed area (base 1990–2009)	continuing trend (base 1990–2009)	continuing trend with higher growth rate
Population[b]	increasing	less increasing	less increasing
Time effects[c]	constant trend	constant trend	constant trend

Notes:[a] In countries with negative annual growth during 1990–2009, the agricultural area was fixed in all three scenarios.[b] Data for population projections were taken from Gaffin et al., 'Downscaling and Geo-Spatial Gridding', 104–23.[c] Time effects were projected by using a linear spline model with different growth rates prior to 1975 and after 1975: $\beta t = \beta_0 + \beta_1 t + \beta_2(t - 1975)^* D_{1975}$, where D_{1975} denotes a dummy variable, which is 1 from 1975 and 0 otherwise; see Richard Schmalensee, Thomas M. Stoker, Ruth A. Judson, 'World Carbon Dioxide Emissions: 1950–2050', *The Review of Economics and Statistics* 80, no. 1 (1998), 15–27.

livestock growth is low, the agricultural area not expanding, and population increase is fast. Further, elasticities and the stock of modern inputs are fixed (no technical improvements and adaptation). In the more optimistic scenario, A1B, livestock growth is higher, agricultural area enlarges and population growth is lower. The third scenario A1B_TT, finally, allows for irrigation and adaptation by enlarged elasticities (representing the 'technology trend' we added to the base scenario). This scenario is quite visionary, but helps to see how small changes in assumptions affect projections. Finally, model 2 (see Table 1) is used to show how agricultural production will develop under rain-fed conditions in the first two scenarios, while model 5 is taken to catch improvements in the third scenario.

As Figure 2 shows, overall production increased from 1961 to 2009. However, simulation results under A2, A1B, and A1B_TT follow different paths. While growth rates are declining in all scenarios over time (by almost half between 2020 and 2050, and respectively one-third in the case of A1B_TT), production is increasing rather slowly in A2, faster in A1B, and quickly with adaptation in A1B_TT. Hence, while an expansion of labor power is not very influential, the intensification of the use of even basic technologies strongly impacts food production and, hence, the chances of food security in SSA regarding average per capita production. This was already mentioned by the IPCC in 2007, which concluded that climate change 'will further shift the focus of food insecurity to sub-Saharan Africa' but also that the impact of the applied 'socio-economic pathways (SRES scenario) on the numbers of people at risk of hunger is significantly greater than the impact of climate change'.[45]

What is more relevant with respect to risk, however, is a comparison we make by splitting the sample into three groups according to the combined average level of income and food supply between 1998 and 2002.[46] This is done to grasp the country's capacity to meet demands by domestic production and imports (the latter certainly being dependent on income). Table 5 reports means of the projection results for these three groups, the first being the group with the highest vulnerability (i.e. with rather low levels of GDP per capita and nutritional energy supply per head and day).

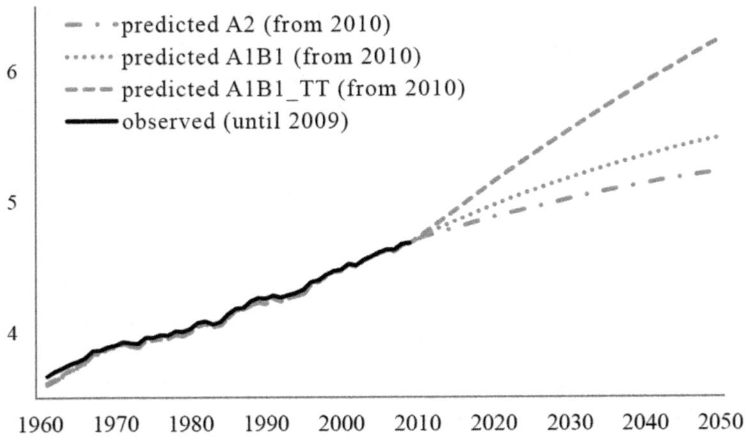

Figure 2. Estimated and predicted agricultural production in sub-Sahara Africa (log value index), 1960–2050.
Source: See relevant text.

Table 5. Average annual growth rates of population and agricultural production in sub-Sahara Africa, 1970–2050

	Population		Production			
	A2	A1B	observed	A2	A1B	A1B_TT
	2010–2050	2010–2050	1970–2009	2010–2050	2010–2050	2010–2050
Country group with ...						
... high vulnerability	2.0	1.6	1.99	1.52	2.18	4.02
... medium vulnerability	1.6	1.3	2.42	1.33	2.10	4.12
... low vulnerability	1.1	0.8	1.83	1.10	1.62	3.56
Total	1.5	1.2	2.11	1.30	1.95	3.90

Source: See relevant text.

All reported numbers are to be taken with great caution because projections contain many uncertainties. But the general trends are clear: the rather pessimistic A2 scenario produces unfavourable results for all groups (the least for the countries with low vulnerability), pointing to an increasing risk in the future; the rather optimistic A1B scenario, on the other hand, points to improvements in all groups and large improvements in the A1B_TT scenario. However, when country-level data is investigated, certain countries of particular risk can be identified. Under the A2 scenario, most countries will likely face a deterioration of food supply on a per capita level and this is more the case, the worse the situation in the country already is. But some regions face this problem also under the A1B scenario, most notably Angola (where only 77% of population growth is met by increased production), Somalia (87%) and Congo Kinshasa (89%).

Two further caveats are necessary. The first is that in all cases distribution is crucial because, while production would in many countries be sufficient to assure food security, malnourishment is prevalent due to unbalanced access. Take the obvious example of Angola, a relatively wealthy country: if the income generated by its rich resources were more equally distributed, current and foreseeable food insecurity problems could completely vanish. The second caveat is that no matter which scenario is applied, technology is absolutely crucial and could result in significant improvements. However, as confirmed by our analysis, this does not mean that it is necessary for Africa to achieve a high-tech agro-industry; in order to achieve food security, it would be sufficient to apply appropriate techniques to improve yields in a low-tech and small-scale farming environment.

Conclusions and outlook

A qualitative summary of our overall results is given in Table 6. The risks are denominated as 'high', 'high-medium', 'medium' and 'low' respectively. The order of countries reflects the risk classes, a preference for empirically well-grounded institutional risk within these classes, and, finally, a comparative assessment of idiosyncratic factors. All countries not mentioned in this list do not come up with more than medium risk in no more than one of the two dimensions.

We draw three conclusions from these results. First, understanding relationships between climate change and violence, very probably transmitted via socio-economic tensions and political moderation, is still insufficient. It is likely, however, that there is a relationship between food security and violence, but it is also likely that this relationship is complex, strongly related to distributional and marginalisation patterns, and nonlinear. Our results also confirm that those countries already vulnerable to violence will most likely remain so and their relative position may deteriorate because of adverse dynamics. Further, communities so marginal that they are virtually invisible and hence not reflected

Table 6. Institutional and environmental risks in sub-Sahara Africa in the twenty-first century

	Institutional risk	Environmental risk
Congo Kinshasa	high	high
Somalia	high-medium[a]	high
Liberia	high	medium
Angola	medium	high
Burundi	high-medium	high-medium
Ethiopia	high-medium	high-medium
Mauritania	high-medium	medium
Sierra Leone	high-medium	medium
Uganda	high-medium	medium
Chad	medium	high-medium
Eritrea	medium	high-medium
Djibouti	high	low
Rwanda	high-medium	low
Nigeria	high-medium	low
Côte d'Ivoire	high-medium	low
Congo Brazzaville	high-medium	low
Burkina Faso	medium	medium
Guinea	medium	medium
Madagascar	low	high-medium
Zambia	low	high-medium

Notes:[a] In the case of Somalia, we lack comparable data with respect to the institutional risk factors; hence, we assessed the risk to be 'high-medium' for obvious reasons (Somalia being a failed state affected by various historical and ongoing inter- and intra-national conflicts).
Source: See relevant text.

in comparative quantitative data (to which this study was restricted) may be particularly affected. When looking at the presence of risk factors from the perspective of potential perpetrators, it is highly likely that the costs of adaptation to climate change are high compared to the political and economic costs of violence and particularly of genocide.

Second, evidence clearly shows that climatic factors are very unlikely to influence the outbreak, intensity and duration of violence directly, while political and socio-economic factors are much more relevant. However, because of the interference of state capacity there is a considerable combined risk. Increasingly adverse effects from climate change, such as a deterioration of food security, and its negative consequences for economic growth and institutional quality, which are also empirically better grounded, may lead to a vicious cycle resulting in certain forms of violence possibly resulting in genocidal episodes. Feasibility also plays a role in genocides, in the sense that the larger the magnitude of political upheaval, the greater the willingness to take 'extreme measures' and the opportunity to seek 'final solutions'.[47]

Third, projections of future agricultural production, in the peripheral environment of SSA the most relevant factor for political stability and economic prosperity directly affected by climate change, are very sensitive to assumptions about technological development, which are consequently crucial. Thus, there is much to be gained even from low-tech improvements. Development, especially in poor countries, would considerably change the actual outcome and peace-building efforts may even completely offset negative effects of climate change. But the complex interaction and the relevance of feasibility also allow for a specific corridor (at a low level) of development, in which the risk of violent outbreaks may increase. Equally, the experiences of past decades raise concerns about a proper utilisation of agricultural technologies and good governance practices in Africa – but not because of African backwardness, but because of the rational decisions of people confronted with specific challenges, existential constraints, and very diverse incentive structures. In the end, the developmental path a country or a society follows, including the degree of violence in general or the occurrence of genocide in particular, is crucially dependent on individual and collective decisions made by the stakeholders involved (including the international community), providing – or denying – the means to combat adverse developments and the effects of climate change.

Hence, further research is necessary; the study of climate change and violence is only just beginning. We would especially welcome comparative case studies of the countries at risk identified in this study and systematic work willing to prudently generalise insights from these studies,[48] as well as further interdisciplinary collaborative quantitative cross-country analysis. While case studies are particularly relevant because they allow a more detailed look at the actual means of transmission of climate change to the probability of conflict and genocide, among those the ones identified by cross-country analysis, the crucial factors moderating transmissions are institutional and not necessarily idiosyncratic.[49] Therefore, we also need institutional approaches to comparatively analyse the effects of climate change on the political and socio-economic sphere and further on the outbreak and continuation of organised and unorganised violence. In particular, we need interdisciplinary studies about threshold conditions and baseline shifts. We also need studies at the largest systemic level to understand how the international system and the global economy affect climate change and the means of adaptation and how these dynamics interact with local patterns.[50] Finally, better data are welcome in all these fields, especially more decentralised and nuanced data on institutions of land-use and the settlement of conflict.

Overall, when facing climate change, violence is not a first best solution at the societal level, but it may be rational from a perpetrator perspective. Further, insufficient institutions are certainly the more serious problem to tackle when striving to prevent genocide. But climate change has important and mutual interconnections with bad institutions. Of particular relevance are connections to marginalisation and vulnerability and the organisation of conflict regulation at the local level when previously marginal groups are affected.[51] This article has contributed to the identification of countries at risk. Although some of these are 'usual suspects' (like in the Great Lakes and the Horn of Africa region), the picture has certainly become clearer and more nuanced with respect to the degree of risk. But for successful prevention to occur, research, policy and activism will have to join forces to identify the places and groups at risk and address these risks by, for example, raising the costs of violence, improving dispute settlement and identifying viable alternatives. The task is mighty indeed, but the benefits certainly outweigh the costs.

Acknowledgements

We thank Alvaro Caldazilla (Kiel Institute for the World Economy) for improving the quality of projections, Ian Harris (East Anglia University) for providing updated climate data, the participants of a session of the 2011 annual meeting of the Austrian Economic Association in Graz, the participants of a session of the 2013 Jan Tinbergen Peace Science Conference in Milan, and an anonymous referee for valuable comments on earlier versions of this article.

Notes

1. See Nicolas Stern, *Stern Review on the Economics of Climate Change* (London: HM Treasury Cabinet Office, 2006); and IPCC, *Climate Change 2007: Fourth Assessment Report* (Geneva: Intergovernmental Panel on Climate Change, 2007).
2. Jürgen Zimmerer, 'From the Editors: Environmental Genocide? Climate Change, Mass Violence and the Question of Ideology', *Journal of Genocide Research* 9, no. 3 (2007): 350.
3. UNDP, *Human Development Report 2007/2008: Fighting Climate Change: Human Solidarity in a Divided World* (New York: United Nations Development Programme, 2007), 90.
4. Charles H. Anderton, 'Choosing Genocide: Economic Perspectives on the Disturbing Rationality of Race Murder', *Defence and Peace Economics* 21, nos 5–6 (2010): 459–86.
5. See for example Jürgen Scheffran, Michael Brzoska, Jasmin Kominek, P. Michael Link, Janpeter Schilling, 'Climate Change and Violent Conflict', *Science* 336 (18 May 2012): 870,

for a schematic view of the nexus between the climate system, natural resources, human security and societal stability; David D. Zhang, Harry F. Lee, Cong Wang, Baosheng Li, Qing Pei, Jane Zhang and Yulun An, 'The Causality Analysis of Climate Change and Large-Scale Human Crisis', *Proceedings of the National Academy of Sciences* 108, no. 42 (2011): 17298, for a general model of the relations between climate change and economy, society and human ecology; and Halvard Buhaug, Nils Petter Gleditsch and Ole Magnus Theisen, *Implications of Climate Change for Armed Conflict* (Washington, DC: World Bank, 2008), for the multifaceted relationship between adverse climate change and the increased risk of armed conflict.

6. Nils Petter Gleditsch, 'Whither the Weather? Climate Change and Conflict', *Journal of Peace Research* 49, no. 1 (2012): 6.
7. This is also the essential argument in UNDP, *Human Development Report 2007/2008*, 90–2.
8. See the PITF homepage, especially http://www.systemicpeace.org/inscr/PITFGenoPoliticides2010.xls.
9. See the Genocide Watch homepage, especially http://genocidewatch.net/alerts-2/newalerts/. The Central African Republic has recently been added to the list of emergencies.
10. For the concept of 'democide' see Rudolph J. Rummel, *Death by Government* (New Brunswick, NJ: Transaction Publishers, 1994).
11. See, for example, Matthew Krain, 'State-Sponsored Mass Murder: The Onset and Severity of Genocides and Politicides', *Journal of Conflict Resolution* 41, no. 3 (1997): 331–60, especially on opportunity; Paul Collier and Anke Hoeffler, 'On Economic Causes of Civil War', *Oxford Economic Papers* 50, no. 4 (1998), 563–73, the origin of quantitative civil war studies of the resource curse type; James D. Fearon and David D. Laitin, 'Ethnicity, Insurgency, and Civil War', *American Political Science Review* 97, no. 1 (2003): 75–90, especially (critical) on ethnicity; Joshua W. Busby, Todd G. Smith, Kaiba L. White and Shawn M. Strange, 'Climate Change and Insecurity: Mapping Vulnerability in Africa', *International Security* 37, no. 4 (2013): 132–72, who virtually map multifaceted vulnerabilities on a regional scale in all of Africa.
12. See Paul Collier and Anke Hoeffler, 'Beyond Greed and Grievance: Feasibility and Civil War', *Oxford Economic Papers* 61, no. 1 (2009), 1–27.
13. Barbara Harff, 'No Lessons Learned from the Holocaust? Assessing Risks of Genocide and Political Mass Murder Since 1955', *American Political Science Review* 97, no. 1 (2003): 61.
14. Harff, 'No Lessons Learned', 67. But there is large uncertainty and the 95% confidence interval in the presence of all risk factors ranges from 67 to 98%. However, Harff was also able to 'postdict' episodes (genocides and non-genocides alike) with a remarkable accuracy of 74%.
15. Ibid., 61.
16. See William Easterly, Roberta Gatti and Sergio Kurlat, 'Development, Democracy, and Mass Killings', *Journal of Economic Growth* 11, no. 2 (2006): 129–56.
17. Ibid., 146.
18. See Frank W. Wayman and Atsushi Tago, 'Explaining the Onset of Mass Killing, 1949–87', *Journal of Peace Research* 47, no. 1 (2010): 3–13.
19. Deborah Mayersen and Stephen McLoughlin, 'Risk and Resilience to Mass Atrocities in Africa: A Comparison of Rwanda and Botswana', *Journal of Genocide Research* 13, no. 3 (2011): 247–69.
20. Jon Barnett and W. Neil Adger, 'Climate Change, Human Security, and Violent Conflict', *Political Geography* 26, no. 6 (2007): 644.
21. Ole Magnus Theisen, Helge Holtermann and Halvard Buhaug, 'Climate Wars? Assessing the Claim that Drought Breeds Conflict', *International Security* 36, no. 3 (2011–12): 105.
22. See Zhang et al., 'The Causality Analysis'; also Richard S.J. Tol and Sebastian Wagner, 'Climate Change and Violent Conflict in Europe Over the Last Millennium', *Climatic Change* 99, nos 1–2 (2010): 65–79, found that conflict was more likely in Europe in periods of cooling, but less likely so in the industrial era.
23. See William Cline, *Global Warming and Agriculture: Impact Estimates by Country* (Washington, DC: Center for Global Development and the Peterson Institute for International Economics, 2007).
24. Erik Gartzke, 'Could Climate Change Precipitate Peace?' *Journal of Peace Research* 49, no. 1 (2012): 177–92, argues that industrial development is at the same time promoting climate change but diminishing the likelihood of inter-state war.

25. See Melissa Dell, Benjamin F. Jones and Benjamin A. Olken, *Climate Change and Economic Growth: Evidence From the Last Half Century*, NBER Working Paper No. 14132 (Washington, DC: National Bureau of Economic Research, 2008).
26. See Quan Li and Rafael Reuveny, 'Democracy and Environmental Degradation', *International Studies Quarterly* 50, no. 4 (2006): 935–56.
27. Conor Devitt and Richard S.J. Tol, 'Civil War, Climate Change, and Development: A Scenario Study for Sub-Saharan Africa', *Journal of Peace Research* 49, no. 1 (2012): 141.
28. Vally Koubi, Thomas Bernauer, Anna Kalbhenn and Gabriele Spilker, 'Climate Variability, Economic Growth, and Civil Conflict', *Journal of Peace Research* 49, no. 1 (2012): 117.
29. See Christopher K. Butler and Scott Gates, 'African Range Wars: Climate, Conflict, and Property Rights', *Journal of Peace Research* 49, no. 1 (2012): 23–34.
30. See Cullen S. Hendrix and Idean Salehyan, 'Climate Change, Rainfall, and Social Conflict in Africa', *Journal of Peace Research* 49, no. 1 (2012): 45. Also Edward Miguel, Shanker Satyanath and Ernest Sergenti, 'Economic Shocks and Civil Conflict: An Instrumental Variables Approach', *Journal of Political Economy* 112, no. 4 (2004): 725–53, find comparable nonlinearities.
31. Clionadh Raleigh, 'Political Marginalization, Climate Change, and Conflict in African Sahel States', *International Studies Review* 12, no. 1 (2010): 69.
32. Busby et al., 'Climate Change and Insecurity'.
33. For more details and discussions see Andreas Exenberger and Andreas Pondorfer, *Rain, Temperature and Agricultural Production: The Impact of Climate Change in Sub-Sahara Africa, 1961–2009*, Working Papers in Economics and Statistics, No. 26 (Innsbruck: Faculty of Economics and Statistics, 2011). We estimated the function $\log(Y_{it}) = \beta_0 + \beta_1\log(L_{it}) + \beta_2\log(V_{it}) + \beta_3\log(F_{it}) + \beta_4\log(K_{it}) + \beta_5\log(M_{it}) + \beta_6\log(R_{it}) + \beta_7\log(PRC_{it}) + \beta_8\log(TEMP_{it}) + \beta_9 \log(Drought_{it})$, with $\beta_0 = \log(A)$, i.e. productivity, *i* denoting country and *t* denoting year. *Y* represents the agricultural output (measured as the Food and Agriculture Organization (FAO) index in value terms), while *L, V, F, K, M* and *R* refer to input factors, i.e. labour, livestock, fertiliser, capital, land and irrigation. *PRC, TEMP* and *Drought* denote the climatic factors: precipitation, temperature and incidence of droughts. Considering the fact that there are unobserved country specific and time varying effects, a two-way error component regression model (two-way fixed effects) is chosen for the analysis. The methodology is based on Jess Benhabib and Mark M. Spiegel, 'The Role of Human Capital in Economic Development, Evidence From Aggregate Cross-Country Data', *Journal of Monetary Economics* 34, no. 2 (1994): 143–73, and Robert Solow, 'A Contribution to the Theory of Economic Growth', *Quarterly Journal of Economics* 70, no. 1 (1956): 65–94.
34. All agricultural inputs are taken from the FAOSTAT database (http://faostat.fao.org/), the agricultural output data *Y* are measured in international US$, indexed with relation to a base period 2004–2006. We introduce the variable *R*, which is irrigated area in thousands of hectares, including areas equipped for full and partial control of irrigation, equipped lowland areas, pastures and areas equipped for spate irrigation. Climate data (normalised average annual rainfall and temperature on country level) are taken from the Tyndall Centre for Climate Change Research and Ian Harris from East Anglia University (http://www.cru.uea.ac.uk/~timm/cty/obs/TYN_CY_1_1.html until 2000, Ian Harris from 2001 to 2009). Data about the droughts are taken from the international disaster database (http://www.emdat.be/). For details see Exenberger and Pondorfer, *Rain, Temperature and Agricultural Production*, 10–14.
35. For other examples of aggregate agricultural production functions in Africa, see for example, George Frisvold and Kevin Ingram, 'Sources of Agricultural Productivity Growth and Stagnation in Sub-Saharan Africa', *Agricultural Economics* 1, no. 1 (1995): 51–61; and Salvador Barrios, Bazoumana Ouattara and Eric Strobl, 'The Impact of Climatic Change on Agricultural Production: Is it Different for Africa?', *Food Policy* 33, no. 4 (2008): 287–98. Only the latter also include climate effects.
36. What we call 'genocidal violence' here is a summary of geno-/politicides according to the PITF definition: 'Genocide and politicide events involve the promotion, execution, and/or implied consent of sustained policies by governing elites or their agents [. . .] that result in the deaths of a substantial portion of a communal group or politicized non-communal group. In genocides the victimized groups are defined primarily in terms of their communal (ethnolinguistic, religious) characteristics. In politicides, by contrast, groups are defined primarily in terms of their political opposition to the regime and dominant groups.' Political Instability Task Force,

State Failure Problem Set: Internal Wars and Failures of Governance, 1955–2012. http://www.systemicpeace.org/inscr/PITFProbSetCodebook2012.pdf (accessed 2 May 2013), 14–15.

37. In an interview study of 340 Rwandan households, it is shown that besides male sex and Tutsi ethnicity, also (higher) age and (larger) land ownership were relevant risk factors for being killed. See Philip Verwimp, 'Testing the Double-Genocide Thesis for Central and Southern Rwanda', *Journal of Conflict Resolution* 47, no. 4 (2003): 423–42. The same is shown for Burundi in 1993 by making use of a representative sample of 7520 households. See Tom Bundervoet, 'Livestock, Land and Political Power: The 1993 Killings in Burundi', *Journal of Peace Research* 46, no. 3 (2009): 357–76.
38. From Table 2, only the episode in Burundi, starting in October 1993, had a possible connection to bad weather, not even climate. In October and November 1993, the two first months of the wet season, rainfall fell short by three quarters compared to a 30-year average.
39. This may also be the case in Ethiopia in the 1970s and even more in the 1980s, when repression by the communist regime and secessionist war were also connected to a deterioration of food security.
40. A factor is regarded as being 'present' when the country performance is below the sample median.
41. A comparable approach is used by Devitt and Tol, 'Civil War, Climate Change, and Development', but their story is about civil war and drought risk as well as the effects of civil wars and climate change on economic growth until 2100.
42. See IPCC, *Emission Scenarios. A Special Report of IPCC Working Group III* (Geneva: Intergovernmental Panel on Climate Change, 2000). The A1 scenario family assumes a future world of rapid economic growth and convergence, increased interactions and low population growth (A1B referring to a 'balanced' use of fossil and non-fossil sources of energy). The A2 scenario family assumes a heterogeneous world of self-reliance, in which economic growth is slower, but population growth larger.
43. We use population projections from Stuart R. Gaffin, Cynthia Rosenzweig, Xiaoshi Xing and Greg Yetman, 'Downscaling and Geo-Spatial Gridding of Socio-Economic Projections from the IPCC Special Report on Emissions Scenarios (SRES)', *Global Environmental Change* 14, no. 2 (2004): 104–23, using the same quantitative procedure for all countries in our sample. Concerning agricultural land, we apply FAOSTAT data and constant area for the A2 scenario, a continued trend (with respect to the average growth 1990–2010) for the A1B scenario, and an accelerated trend (1.5 times the average growth 1990–2010) for the A1B_TT scenario. Concerning livestock, we apply a linear trend and assume an increase by 50% until 2050 (with respect to the average stock 2005–2009) for the A2 scenario, and an increase by 150% for the A1B and A1B_TT scenarios. Concerning tractors, fertilisers and irrigation, we assume constant stocks (only in the A1B1_TT scenario, the elasticity of fertiliser and irrigation is constantly increasing to 0.1 in 2050, which is not overly optimistic, because Barrios et al., 'The Impact of Climatic Change', for example, estimated a coefficient of 0.078 in Asian countries for the use of fertilisers in the period 1961–2000).
44. Temperature and rainfall projections are calculated as comparisons of 30-year averages (from 14 years before until 15 years after the respective year) with base period averages (1961–1990), calibrated with Tyndall Centre for Climate Change Research data. Projections are based on Pete D. Falloon and Richard A. Betts, 'The Impact of Climate Change on Global River Flow in HadGEM1 Simulations', *Atmospheric Science Letters* 7, no. 3 (2006): 62–8; Peter A. Stott, Gareth S. Jones, Jason A. Lowe, Peter Thorne, Chris Durman, Timothy C. Johns and Jean-Claude Thelen, 'Transient Climate Simulations With the HadGEM1 Climate Model: Causes of Past Warming and Future Climate Change', *Journal of Climate* 19, no. 12 (2006): 2763–82; Timothy C. Johns et al., 'The New Hadley Centre Climate Model HadGEM1: Evaluation of Coupled Simulations', *Journal of Climate* 19, no. 7 (2006): 1327–53; and G.M. Martin, M.A. Ringer, V.D. Pope, A. Jones, C. Dearden and T.J. Hinton, 'The Physical Properties of the Atmosphere in the New Hadley Centre Global Environmental Model, HadGEM1. Part 1: Model Description and Global Climatology', *Journal of Climate* 19, no. 7 (2006): 1274–301.
45. IPCC, *Climate Change 2007*, WG2, Ch. 5, Section 5.8.1 'Findings and Key Conclusions'; these particular conclusions are made 'with high confidence' in a statistical sense.

46. The data used to construct a multiplicative index are taken from Angus Maddison (GDP) and FAOSTAT (nutritional energy); see http://www.ggdc.net/MADDISON/oriindex.htm and http://faostat.fao.org/ respectively.
47. Harff, 'No Lessons Learned'; Krain, 'State-Sponsored Mass Murder'.
48. Although Sudan is excluded from our analysis and Kenya does not figure prominently because of its relatively positive agricultural outlook, on average, a connected example for a study combining theoretical and empirical work is provided by Jürgen Scheffran, Tobias Ide and Janpeter Schilling, 'Violent Climate or Climate of Violence? Concepts and Relations with Focus on Kenya and Sudan', *The International Journal of Human Rights* 18, no. 3 (2014): 369–90.
49. On the interplay between climate change and culture in the case of the Pacific island of Chuuk, see Rebecca Hofmann, 'Culturecide in Changing Micronesian Climates? About the Unintenionality of Climate Change', *The International Journal of Human Rights* 18, no. 3 (2014): 336–49.
50. See Gregory Kent, 'Crystallisations of the Global Western State in the Era of Climate Change', *The International Journal of Human Rights* 18, no. 3 (2014): 320–35.
51. See Mark Levene and Daniele Conversi, 'Subsistence Societies, Globalisation, Climate Change and Genocide: Discourses of Vulnerability and Resilience', *The International Journal of Human Rights* 18, no. 3 (2014): 281–97.

Violent climate or climate of violence? Concepts and relations with focus on Kenya and Sudan

Jürgen Scheffran[a], Tobias Ide[a,b] and Janpeter Schilling[a,c]

[a]Research Group Climate Change and Security, Institute of Geography, University of Hamburg, Germany; [b]School of Integrated Climate System Sciences (SICSS), University of Hamburg, Germany; [c]International Alert, London, UK

Addressing deficits of current research on the link between climate change and violent conflict, this article aims to contribute to a more systematic understanding of the violence concept in the context of environmental change. We present a theoretical framework and potential pathways between climate change and violence and an agent-based approach to assess the interplay between capabilities and motivations for violence and the conditions for conflicting or cooperative interactions. Acting as a 'threat multiplier', climate change could exceed adaptive capacities and undermine the livelihoods of communities. In the most affected regions, the erosion of social order and state failure as well as already ongoing violent conflicts could be aggravated, leading to a spiral of violence that further dissolves societal structures. Against this background we analyse case studies in Kenya and Sudan, focusing on factors driving or preventing a spiral of violence. While interpastoral conflicts in north-western Kenya result in limited numbers of casualties, the Darfur conflict has been shaped by the civil war in Sudan, involving the government, rebel forces and militias, causing significant loss of lives and destruction. The impact of climate change is less direct in Sudan than in Kenya. To avoid a spiral of violence, in both cases it is essential to reduce socio-economical marginalisation, develop resource-sharing mechanisms and restrain access to arms as part of long-term strategies for a sustainable and peaceful intervention to contain the adverse impacts of climate change.

1. Introduction

In recent years there has been increasing debate on the security risks of anthropogenic climate change.[1] Freshwater scarcity, food insecurity, extreme weather events and environmental migration[2] are being discussed as key processes that are affected by climate change and potentially relevant to the onset of violent conflict.[3] However, there is no scientific consensus yet. Climatic changes are found to contribute to violent conflicts in some cases while other studies do not support such a conclusion.[4] Part of this heterogeneity can be explained by the different concepts of violence that are applied.

In this article, we first seek to clarify conceptual issues by defining and comparing the concepts of violence used in studies on climate change and violent conflict. Then we explore a theoretical framework and potential pathways from climate change to violent conflict. Here, we discuss how climate change can feed into a spiral of violence in which the use of violence provokes more violence. Against this background, we focus on Kenya and Sudan to serve as practical examples of the role of climate change in situations of reciprocal use of violence and how such violence could be prevented. The last section combines the key findings of the previous sections and considers how a spiral of violence fuelled by climate change can be avoided.

2. Concepts of violence in research on climate change and violent conflict

2.1 *Basic concepts*

Following Bonacker, a conflict can be defined as a situation in which the incompatible expectations of at least two actors meet.[5] If differences and tensions cannot be overcome, conflicting actions generate mutual losses and escalate, pushing conflict parties towards extreme actions, such as the use of violence. While often causing severe losses, conflicts can also be a constructive force for social change.[6] In cases of undesirable conflicts, conflict resolution may help to reduce the incompatibility of the actors' expectations to pave the way for de-escalation and even for cooperation. Conflict resolution requires adaptation towards common positions and mutually beneficial actions that stabilise the interaction. Whether this succeeds depends on several factors, including the governance capacity of societies to manage conflicts.

Like conflict, violence has a broad range of meanings. The World Health Organization defines violence as 'the intentional use of physical force or power, threatened or actual, against oneself, another person, or against a group or community, that either results in or has a high likelihood of resulting in injury, death, psychological harm, maldevelopment or deprivation.'[7] This definition associates the intention to apply violence with the capability to use a type of force requiring some kind of arms, thus inherently inferring a destructive mechanism. Common in conflict studies is the definition of armed conflict 'as a contested incompatibility that concerns government or territory or both where the use of armed force between two parties results in at least 25 battle-related deaths. Of these two parties, at least one is the government of a state.'[8] An armed conflict is a particular form of collective violence committed in an institutionalised manner by larger groups or by states. Among the most devastating forms of collective violence is genocide, which, according to the 1948 Genocide Convention, is defined as intentional acts committed 'to destroy, in whole or in part, a national, ethnic, racial or religious group'.[9] In contrast to behavioural violence involving an acting subject, structural violence as suggested by Galtung is indirect and built into the societal structure, affecting the distribution of resources and life chances.[10] The post-liberal tradition refers to structural violence in broader historic contexts, ranging from the contest for land in settler colonialism[11] to today's extractive industries[12] which may have consequences similar to those of genocide.

2.2 *Violent conflict and environmental change*

Among the expected consequences of climate change are renewable resource scarcity (such as water, land or food) and migration in response to more frequent or intense natural disasters. These are factors identified as potentially relevant for conflict.[13] The large majority of studies on a possible climate–conflict nexus focus on violent conflicts within states, between groups, and in the Global South. This is not surprising given the fact that violent conflicts between states are less likely to be influenced by environmental factors than intra-state ones.[14] Similarly, many locations in the Global South are vulnerable to climate change and especially prone to inter-group violence. Traditionally peace and conflict research focuses less on individual-level violence than on group-level conflicts. In addition, policy makers, funding agencies and the wider public tend to pay more attention to intergroup violence (at least in the global south) than to domestic violence or robbery rates. One possible explanation is that individual-level violence, in contrast to many forms of intergroup violence, is not expected to cause large-scale conflict or migration towards the global north.[15]

Studies on the link between environmental or climate change and violent conflict often do not explicitly define concepts of conflict or violence.[16] Other authors provide lists of phenomena they subsume under the term violent conflict, but give no explanation about how these phenomena are defined or related to each other.[17]

In recent years, there has been a trend in conflict studies in general and especially in the research on violence induced by environmental or climate change to conduct large-N studies with statistical methods. Every data set used by this strand of research is usually delivered with a codebook specifying which events of violent conflict were included based on which criteria. We reviewed 45 definitions of violence used in 39 large-N studies[18] on the connection between climatic changes, resource scarcity or natural disasters and violence in order to answer the following questions: What does violence mean? Who uses violence? Against whom is violence used? Is violence reciprocal or one-sided? The results are summarised in Table 1.

As Table 1 shows, most studies investigating a possible link between environmental or climate change and violent conflict define violence as the intended, direct and organised killing of people, usually with the help of arms and in battles (38 usages). Only three studies integrate the destruction of property into their definition of violence, while no study uses the sole hurting of humans as a violence threshold or includes structural forms of violence. The majority of the studies examined require at least one conflict party to be a state for a violent event or conflict to be recognised (29 usages). We found only nine definitions referring to violence used by organised groups and seven definitions referring to (not well-organised) social violence. Finally, violence was most often understood as a reciprocal process (35 usages), while only ten studies allowed for violence to be a one-sided process (such as massacres or genocides). The standard definition of violence in the quantitative research on climate change and violent conflict can thus be summed up as follows: *Violence is the direct, intended and reciprocal killing of people taking place between state forces and an organised non-governmental group.* This

Table 1. Overview of concepts of violence as used in large-N studies (for references see Appendix I)

Aspect	Specification	Number of usages (n=45)
Meaning of violence	Intended killing of people (by arms)	38
	Intended destruction of substantial property	3
	Not clearly stated	5
Actors of violence	State vs. armed group/civilians[a]	29
	Organised group vs. organised group/civilians	9
	Social group vs. social group	7
Direction of violence	Reciprocal	35
	One-sided	10

Notes: Many of these specifications also include violence between two or more states, but this kind of violent conflict is hardly influenced by climate change.

definition is strongly influenced by the definition of violence of the UCDP/PRIO armed conflict data set,[19] which was used 21 times in the studies examined, equal to nearly 50% of all usages.

The identified standard definition seems to also fit many qualitative studies. Consider, for instance, Colin Kahl, who defines his major dependent variable, civil strife, 'as large scale, sustained, and organized violent conflict within a country. This includes revolution, rebellion, insurgency, civil and ethnic war, and sustained campaigns of terrorism.'[20]

However, while the quantitative literature so far is mainly concerned with instances of violence involving at least one state actor, there is a growing awareness that climate change is more likely to lead to small-scale communal violence without state involvement. Thus, recent case studies[21] and some large-N investigations[22] also include violence between organised armed or social groups. Additionally, some qualitative studies pay attention to forms of violence which do not necessarily result in killing.[23] Nonetheless, more research is necessary to address the discrepancy between the statement that climate change will more likely be associated with small-scale violent events and the assertion that good data are available mainly for more visible events such as armed conflicts involving the state.

3. Conceptual framework of climate change and violent conflict

To address the complex interlinkages between climate change and different forms of violence, particularly violent conflict, we now provide and discuss a conceptual framework to assess the environmental and societal conditions for the use of violence.

3.1 *Environmental degradation, human security and societal stability*

Given the ambiguities, uncertainties and limitations of current research on the relationship between climate change and violent conflict, there is a distinct need for an integrative assessment that supplements quantitative empirical analysis and qualitative case studies. A promising approach is the analysis of possible complex pathways and feedbacks between climate change, natural resources, human security and societal stability.[24] Such a systemic framework not only addresses the direct interaction between the climate system and societal stability, including violent conflict, but also the impacts on natural resources and human security, which may in turn affect society and the likelihood of violent actions.

The impact of environmental change on human communities depends on their vulnerability, being a consequence of the exposure and sensitivity to climate change as well as the adaptive capacity that is specific for each region.[25] These factors affect human security, defined by Barnett and Adger as 'the condition where people and communities have the capacity to manage stresses to their needs, rights, and values'.[26] The causes and factors that affect human security and determine the ability to adapt to and mitigate the effects of climate change are unevenly distributed. People who are already poor, politically marginalised or affected by violent conflicts may be more sensitive and less able to adapt to climate change-related disasters and resource scarcities that contribute to different dimensions of human insecurity such as hunger, poverty and disease. Generally, agricultural societies[27] with low levels of economic income are more dependent on natural resources and ecosystem services than industrial societies with high income. Those can buy food or clean water or build appropriate shelters to protect citizens from economic and social risks caused by environmental change.[28] These asymmetries and inequalities of human insecurity are partly associated with the legacy of colonialism, capitalist globalisation and neoliberal structural reforms.[29]

If environmental change severely affects human security, this could lead to socio-economic stress, societal instability and violent conflict. When societal structures, such as states or traditional conflict mediation and resource sharing practices, are undermined, they may lose credibility and support from citizens, become weak and remain unable to maintain social order. Climate risks could multiply other societal problems that together could overwhelm the problem-solving capacity of societies, disrupt governments and trigger societal instability events, including a smaller number of large-scale events (such as civil wars) and a larger number of small-scale events (protests, riots, intergroup and individual violence). Climate change is, therefore, often predicted to be a 'threat multiplier' rather than a direct cause of violent conflict.[30] Due to non-linear effects, an increase in global temperature above a certain threshold (such as 2°C) may result in disproportionate impacts and abrupt climate change. Tipping elements and cascading sequences could, in the long-term, overwhelm the adaptive capacity even of wealthy nations.

3.2 *Conditions and dynamics of violent action*

In the following section, we describe a conceptual framework for the use of force and violence in terms of three factors that are of basic relevance for human action: motivation, capability and the natural and societal context (Figure 1). We use a structural-systemic approach, being aware that each of the mechanisms is affected by subjective perspectives and societal discourses and embedded into power relations:

(1) Acts of behavioural violence require the capability to cause damage and destruction reliant on particular tools, such as arms or soldiers. The impacts of force also depend on the vulnerability of those affected, which is a function of defence capabilities and the ability to recover after the use of violence. However, some insecurity always remains, for instance, from unexpected surprise attacks such as suicide bombings that can overwhelm even powerful defences.

(2) Motivation is represented by the reasons and incentives why actors use violence to pursue particular goals or prevent certain acts of others, which depends both on expected gains, costs and potential losses/risks (e.g. by individual or collective violence) and on prevalent concepts of enemies.

(3) The natural and societal context provides the environment that supports, contains or transforms the use of individual or collective violence. In an environment of insecurity and violence, people feel threatened and pressed to acquire and use force to protect themselves and their property from the threat of others, possibly leading to a self-stabilising collective 'climate of violence'. Whether such an environment is supportive of cooperative, confrontative or even violent behaviour depends on the political-economic conditions, dominant identity configurations, class and power structures and ethnic make ups of the respective societies and how these filter environmental stresses, affect vulnerabilities, restrain human capacities and drive political responses.

Once critical thresholds of insecurity and violence have passed, a self-enforcing spiral of violence may emerge when violent acts provoke more violent acts. Conflict dynamics increasingly shape societal context, affecting both motivation (shifting the original

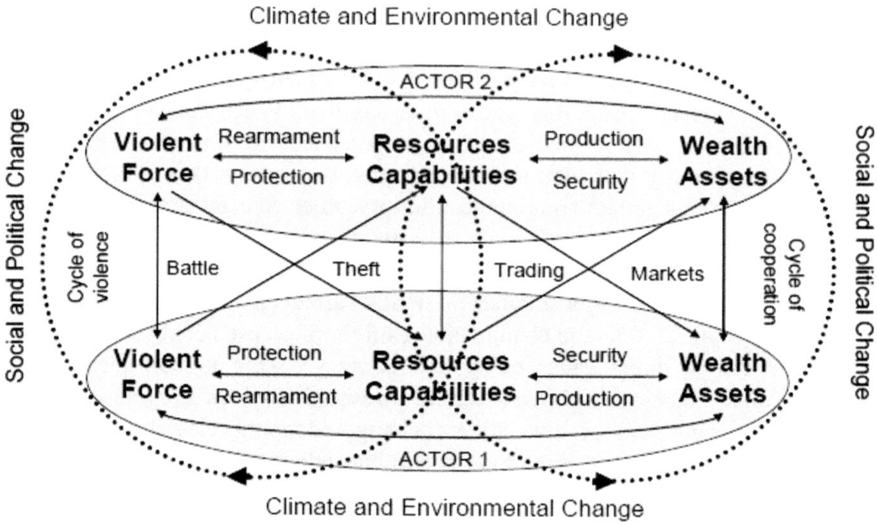

Figure 1. Framework of interaction between two actors, with key variables and pathways indicating cycles of violence and cycles of cooperation
Source: The authors.

incentives towards survival and countering opponents) and capability (increasing losses and needs for violent force). Violence can transform[31] (such as from intercommunal violent conflict to insurgencies and even interstate wars) and spread to neighbouring states or regions,[32] for example, through (cross-border) migration, ethnic links, natural resource flows, black markets or arms exports. Other features also become crucially important for escalation dynamics, such as rough terrain or oil reserves for insurgencies, gender norms for household violence or cultures of violence for armed struggle between different communities. Particularly prone to the development of spirals of violence are (parts of) societies in transition or on the edge of instability, such as fragile and failing states with social fragmentation, weak governance structures and inadequate management capacities.[33] These states cannot guarantee core functions of government, such as law and public order, welfare, participation, a monopoly on the use of force and basic public services such as infrastructure, health and education. Consequently, the power vacuum is filled by actors such as international troops (for example, United Nations peacekeepers), private security companies, terrorist groups, warlords or local militias.

Key variables and pathways of interaction between two actors are presented in Figure 1. Here, the resources and capabilities of each actor may be applied either for productive purposes (the creation of wealth assets) or for destructive purposes (the building and use of violent force). While the former is associated with cooperative forms of interaction, including sharing and trading of goods on markets, conflictive and violent forms involve theft, raiding and battle. While actors can choose to switch from productive to destructive purposes (and vice versa) according to their motivations, collective mechanisms could drive actors into self-enforcing cycles of cooperation or cycles of violence. Insecurity tends to feed a spiral of violence where resources and capabilities are increasingly transferred into rearmament and use of violent force for attack and defence in battle, while an environment of peace, security and disarmament supports a cycle of cooperation. Changes in the natural and social environment may affect this interaction and trigger a transition between cycles of violence and cooperation.

3.3 *Possible linkages between natural resources, environmental change and violence*

After introducing the systemic framework of climate-society interaction and the agent-based framework of violent interaction, in the following section we discuss multiple and complex potential linkages connecting violence to environmental change (in particular climate change and natural resources):

(1) Since human beings need resources to live, produce and consume, resource scarcity can increase the motivation to acquire or defend resources by the use of violence, individually or collectively.[34] This can in turn contribute to resource destruction. When people are forced to migrate, this can cause conflicts due to scarce resources or cultural differences in receiving areas.[35]

(2) Resource scarcity and natural disasters can undermine state capacities and reduce state legitimation (for example, due to decreasing taxes, inadequate state support, loss of public order and infrastructure), thus supporting the conditions for state failure and insecurity. The erosion of the social order and state failure could be aggravated, leading to a spiral of corruption, crime and violence that could involve non-state actors such as private security companies, terrorist groups and warlords.

(3) The abundance of valuable natural resources may provide an incentive to acquire these resources violently. Access to resources may in turn raise the capabilities of actors to use violence, such as an increased ability to buy weapons or pay soldiers.[36] It is often the unrestrained resource demands of wealthy communities that enable this connection between natural resources and behavioural violence.

(4) Violence can serve as an instrument of political elites to enforce environmentally harmful policies. Under certain circumstances, this is countered by violent resistance of affected people, possibly leading to conflict escalation or a spiral of violence. The case of copper mining in Bougainville, Papua New Guinea, is illustrative here.[37]

(5) Climate change, as expressed by harsher climatic conditions and intense natural disasters, can disrupt established human–environment relationships to overwhelm the adaptive capacities of groups and threaten the livelihood of entire communities. This may endanger human security and aggravate already challenging living conditions that can be interpreted as a form of structural violence caused by CO_2-intensive lifestyles.

(6) Strategies against climate change may cause conflicts leading to violent conditions. Examples are food insecurity caused by bioenergy, nuclear proliferation triggered by nuclear energy or the struggle over the intentional manipulation of the climate system by climate engineering.

(7) Finally, environmental change may not only contribute to structural violence but can also make societies more vulnerable to different forms of behavioural violence. Violence in turn can make societies more vulnerable to environmental change, leading to a trap from which escape is difficult. This double exposure of violence and environmental hazard can be observed in a number of countries with low human development.[38]

These multiple linkages between environmental change and violence may interact with each other in complex ways. Violent conflict in Afghanistan, for example, was not caused by environmental factors. However, insecurity about future resource access in times of war and land mines covering farming land led to a rapid overexploitation of natural pistachio woods, which in turn caused deforestation, soil erosion and water scarcity. Today, the very environmental change (and livelihood insecurity accompanied by it) caused by the war make it harder to end the violent conflict.[39]

The seven violent pathways suggested above and their combination could become existential threats to both ecosystems and social systems. A large-scale destruction of the natural environment could result in 'ecocide', which entails extensive or lasting harm to the natural environment on a massive scale.[40] This may become a precursor to the elimination of living spaces, cultures, economies and even inhabitants of communities depending on the environment. Thus, very large-scale environmental change may have consequences similar to genocide, although this is usually not an intended act but is an inadvertent (although often consciously accepted) side effect. Whether justified or not, there are concerns that 'ecocide and genocide may mutually enforce each other'.[41] Ecocide can indirectly cause genocide not only by inducing behavioural violence but also through structural violence, destroying the extra-human material environment and thus undermining the foundations of societies and communities. There are several historical cases describing societal collapse brought on by man-made ecological destruction.[42] Recent studies take a broader view on crimes and harms affecting the environment, human and non-human life and the whole planet,[43] including industrial pollution and illegal disposal of toxic waste, the impact of military operations on landscapes, water supply, air quality, living organisms and, more generally, the exploitation of both the earth's resources and degradation of the environment; these last two elements in particular might have genocidal consequences, such as large famines or mass refugee movements. Rwanda is frequently mentioned as an example showing a link between environmental change and genocide.[44] Some studies focus on cases of 'industrial genocide' that have severe ecological and human consequences. An example are the tar sands in northern Alberta, Canada, that pollute the environment and affect the lives of many indigenous groups by undermining their ability to hunt, trap and fish.[45] A new debate is emerging on responses for more effective and appropriate models of justice and law that consider 'ecocide' as an international crime against peace and human rights.[46]

3.4 *Transformation to prevent a spiral of violence*

Whether climate change will fuel a cycle of environmental change, under-development and violent conflict, or whether a cycle of security, sustainability and cooperation can be achieved depends on human and societal responses to climate change. Strategies that reduce the vulnerability to climate change, address the underlying causes of conflict and tackle the factors encouraging conflict escalation can be effective in preventing violence. To reduce the risk of climate change-related violence, transformations to stable cooperative human–environment and human–human interactions are necessary. In order to succeed in this venture, we must consider how fast and effective these transformations are compared to the speed and intensity of climate change. These transformations offer opportunities for constructive social change and even the resolution of conflicts if new problems can be avoided, for example maladaptation.[47] Strategies providing communities with additional viability and resilience can strengthen the social capability of people in their effective, creative and collective efforts to handle the problems of climate change and other problems they confront.[48] While in the end it is always the individual who chooses to use or not to use violence,[49] the avoidance of violence related to climate change can be located at several levels, including individual actors, local and global communities and state institutions.

4. Regional case studies in Africa

To investigate the conceptual framework of climate change and violence, we will discuss two case studies in Africa, each highlighting a different focus. When addressing north-western Kenya, we will discuss the impact of precipitation patterns on violent raiding. Darfur (Sudan) has been described as one of the first areas for climate conflicts but the environment is only one of many roots of violence in the region.

4.1 *Spiral of violence in north-western Kenya*

4.1.1 *Background*

North-western Kenya is a region characterised by socio-economic marginalisation, poverty and harsh climatic conditions.[50] The most viable livelihood here is pastoralism. Pastoralists from a variety of ethnic backgrounds move with their herds in search of water and pasture. Both resources and livestock are subject to conflicts between pastoral groups.[51] Neither resource conflicts nor the occurrence of drought are new phenomena in north-western Kenya.[52] However, in recent decades, some conflicts have escalated while the drought frequency has increased, possibly because of climate change.[53] We use violent conflicts between the ethnic groups of the Turkana of Kenya and the Pokot of Kenya and Uganda to illustrate how drought in combination with socio-economic marginalisation has fuelled a spiral of violence.

Violence in this case is used to gain or secure control over water, pasture and livestock. It is mostly conducted in the form of violent livestock thefts, called ‘raids’ which are executed reciprocally by the male youth.[54] Periodically, the Kenyan government interferes with the raiding cycle between the Turkana and Pokot through disarmament or punishment campaigns.[55] In several instances, these campaigns have led to violence between governmental forces and the concerned pastoral groups. In combination with the increased drought-related resource scarcity, the hostilities create a strong perception of insecurity.[56]

4.1.2 *Spiral of violence*

Figure 2 shows that insecurity can be seen as the starting point of the spiral of violence. Abandonment of cooperative resource sharing in the form of reciprocal grazing arrangements which increases resource scarcity and distrust between communities, for example,

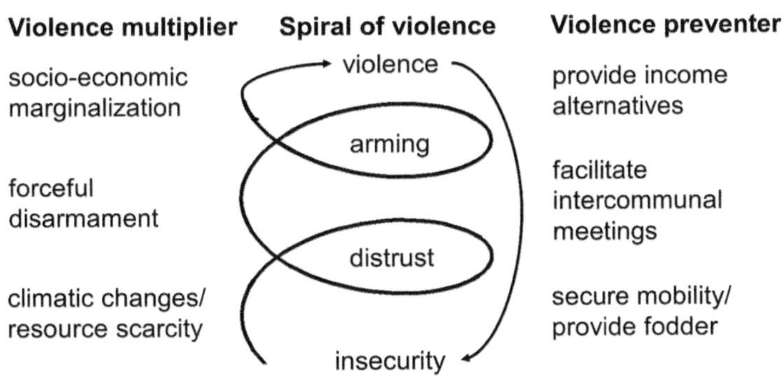

Figure 2. Spiral of violence in north-western Kenya
Source: The authors.

are key ingredients of the next loop.[57] Both insecurity and distrust increase the need for protection, which the communities seek to satisfy through arming.[58] Full automatic weapons, primarily AK-47s, are widely available, inexpensive and powerful.[59] They are used by their owners to defend themselves or to acquire livestock or related resources.[60] Hence, the groups have both the motivation and capability needed to execute violence.

To date, the government of Kenya has had only limited success in preventing or decreasing intercommunal violence. In only 13.4% of the raids conducted between 2006 and 2009 in Turkana, the government has taken action[61] and 8.2% of animals stolen were recovered. The limited number of governmental security forces in north-western Kenya face multiple challenges. Poor road and communication infrastructure, limited vehicles and out-dated equipment reduce the chance to prosecute raiders who are highly mobile and familiar with the remote and partly rough terrain. Alternatives to raiding are limited by high levels of poverty and socio-economic marginalisation.[62] Some authors have pointed out that parts of the pastoral production system in West Pokot are integrated into the wider market economy through 'commercialised raids' in which livestock is stolen and then sold on the market instead of kept to restock own herds.[63] However, in Turkana there is little indication of commercialised raiding.[64] Instead, the region has started to experience the externalities of the capitalist mode of production as significant oil resources are now being exploited in Turkana.[65] Environmental degradation, loss of access to land, and interruption of migration routes by pipelines are attributes of the 'resource curse' that oil exploitation brings to pastoral areas.[66]

Apart from these socio-economic developments, water and pasture availability has decreased mainly due to the higher frequency of prolonged droughts.[67] Additionally, rainfall variability has increased; this in turn decreases the reliability and predictability of water and pasture resources.[68] Climate projections for East Africa indicate that the trend of strong warming and increased precipitation variability will continue.[69] Then, the multiplier factors listed on the left side of Figure 2 may lead to a renewed start of the spiral of violence.

4.1.3 *Preventing and interrupting the spiral of violence*

The right side of Figure 2 shows options to prevent a 'climate of violence' in north-western Kenya. As identified above, the key causes of pastoral insecurity are hostile interactions with other pastoral groups and resource scarcity varying with climatic conditions.[70] Accordingly, structural factors play a role. Commercialisation of livestock raiding partly drives the motivation of the Pokot to use violence as a means to increase wealth, while oil exploitation is likely to aggravate resource scarcity and limit mobility for the Turkana. This is critical as mobility is the central element of pastoralism. Thus far, free and safe movement of pastoralists within Kenya and across international borders is not guaranteed. Turkana pastoralists have reported they are being stopped and harassed by Ugandan security forces when they enter the country in search of water and pasture.[71] Here, international agreements between Kenya and its northern neighbours Uganda, South Sudan, Ethiopia and Somalia are needed. Attempts in this direction have been made but measures have not yet been implemented.[72] Secure cross-border movement and a limitation of the oil exploitation would take pressure off the inner Kenyan conflict between the Turkana and the Pokot but it would most likely not solve conflict completely. To decrease conflicts regarding water and pasture during periods of resource scarcity, especially during droughts, governmental and non-governmental organisations (NGOs) could provide not only food relief, but also fodder subsidies and veterinarian services in order to strengthen the resilience of pastoral communities and buffer drought-related shocks.

When distrust between communities has already built up, it is most promising to undertake measures geared to improve relationships. Intercommunal (peace) meetings can be useful here. For instance, a peace treaty between the Turkana and the Matheniko of Uganda has worked for the past 40 years. However, peace meetings between the Turkana and the Pokot mostly facilitated by NGOs show little success so far. One reason for the limited success may be that mostly elders and chiefs attend the peace meetings while young adults as actual conflict actors are neglected.[73] Providing other income options can decrease motivation to engage in raiding and violence. Investments in educational infrastructure would offer young persons the potential to find a job in other economic sectors, although they are missing as a labour force in pastoralism.

In summary, the spiral of violence between the Turkana and the Pokot originates from a strong perception of insecurity in both communities, fuelled by a variety of human and climate drivers. Regardless, several potential violence preventers exist. If properly and fully utilised, these preventers have the potential to break and prevent the spiral of violence.

4.2 *Violent climate in Darfur, Sudan*

4.2.1 *Background*

For several decades Sudan has been devastated by political instability and violent conflict, aggravated by national power games, regional fights and global geopolitics (most related to oil and terrorism).[74] At the domestic level, major drivers of conflict have been attempts to control power and resources by an Arab ruling elite and the marginalisation and exclusion of peripheral regions such as Darfur which contribute to national disintegration and secession. The second civil war between the north and the south of Sudan started in 1983, while violence in Darfur escalated in 2003.

Natural resource problems and environmental change have contributed to both conflicts. Competition over oil and gas resources was a driving force in the north–south civil war and in the violent conflict in Darfur and remains a source of political tension today. The cold war between the US and the Soviet Union played a major role as does now the competition between the US and China over oil, notably regarding the production of oil pipelines and investments in South Sudan.

Environmental degradation caused by large-scale resource-related development projects, such as the Jonglei canal, also contributed to tensions heightening the north–south conflict. The continued degradation of rangelands in the northern Sahel zone, affected by severe droughts in the 1980s, increases competition over resources between pastoralist groups and sedentary farmers. The Darfur region has been controversially discussed as an example of how climate impacts may or may not interfere with security. The environmental drivers and preventers in the resulting spiral of violence in Darfur are discussed in the following (Figure 3).

4.2.2 *Drivers in the spiral of violence*

The conflict nexus in Darfur is highly complex, with a number of mutually enforcing factors and destabilising interactions. In recent decades, average temperature increased, rainfall significantly declined, and the number of droughts increased, possibly shifting the boundary between semi-desert and desert.[75] These trends are likely to continue and combine with

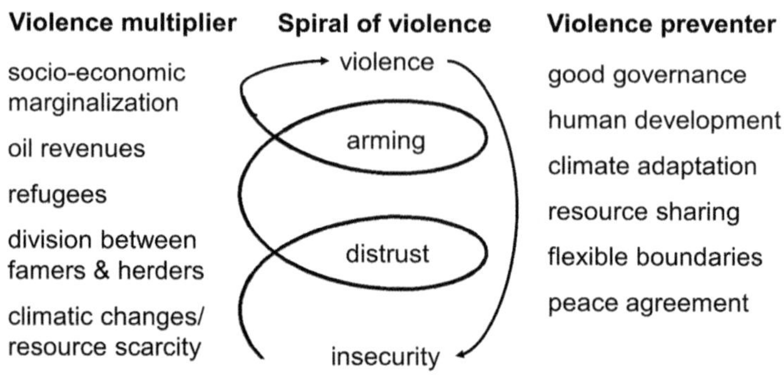

Figure 3. Spiral of violence in Darfur, Sudan
Source: The authors.

population pressure, unsustainable land exploitations and deforestation to diminish agricultural productivity, aggravate food insecurity and expand diseases like malaria. Although Sudan has substantial freshwater resources, including almost two-thirds of the Nile basin and considerable groundwater reserves, mismanagement and inadequate investments have resulted in the decline of access to safe drinking water, contributing to poor health and sanitation conditions.[76]

Increasingly, nomadic groups from the north have moved southward in the dry season in search of water and grazing for their cattle, adding to the pressure on scarce resources in the areas of destination and increasing tension with settled farmers.[77] The expansion of rain-fed mechanised farming systems deprived nomadic herders of much of their traditional migration routes and expropriated traditional smallholder farmers from their land, leading to severe conflicts and boundary disputes.[78] Traditionally, differences over land use were resolved by negotiation among tribal leaders; the government's attempt to establish new administrative structures has recently weakened the established tribal system.[79]

This unstable nexus overlaps with the conflict between the Khartoum government, supported by Janjaweed and other militias, and the Sudan People's Liberation Army/Movement (SPLA/M). While the government headed for the exploitation of the rich oil resources of Darfur, the rebel forces protested against the inadequate compensation and socio-economic marginalisation of the region in general. The government began to frame the conflict as a struggle between 'African' settled farmers and 'Arabic' nomadic pastoralists, in spite of the fact that both were indeed very heterogeneous groups. Identifying groups in this manner transformed pre-existing, resource-related tensions into (perceptions of) an identity conflict, thus allying with (and supporting) the pastoralists in order to eliminate the supposed alliance of rebels and farmers.[80] Combining these two conflicts increased the spiral of violence in Darfur, which was further fuelled by the flow of arms into the region. The resulting acts of evictions, killings, destructions and other abuses further increased prevailing insecurity and created a general climate of violence involving genocidal acts. The high population density in Darfur's refugee camps, in turn, created further environmental and health problems, such as infectious diseases and deforestation around the camps due to excessive fuelwood collection.[81] However, the role of climate change as a threat multiplier or even as a cause of the war in Darfur remains disputed. While some conclude that Darfur is a 'tragic example of the social breakdown that can result from ecological collapse',[82] other studies warn of the 'danger of oversimplifying Darfur'[83] and criticise that blaming climate change could distract from the political causes of the conflict.[84] Altogether, conflicts are caused by a multiplicity of factors, where environmental and political factors represent different and sometimes interconnected dimensions.

4.2.3 *Preventing and interrupting the spiral of violence*

Pending future avoidance of military action between Sudan and South Sudan, one lesson that can be drawn from the Darfur case is that climate change adaptation should be integrated into development and post-conflict reconstruction efforts. Possible measures include a diversification of the (predominately agricultural) economy, improved water resource management or a more effective early warning system for disasters like droughts.[85] A key challenge in this context is to develop a stable system for rural land tenure which is adaptive to changing environmental conditions, addressing long-standing boundary disputes between communities of farmers and pastoralists.

Accountability and transparency of private sector and government decision-making have to be increased, especially concerning large investments in the energy, water and agricultural sectors. Resource use needs to generate considerable benefits in the region where it takes place, especially with regard to livelihood security, health and education.[86] Weak institutional structures for environmental governance could be improved to overcome the deficits in funds, staff and the lack of enforcement of environmental legislation, building on the Comprehensive Peace Agreement and the Sudan Multi-Donor Trust Funds responsible for coordinating the reconstruction and development needs of Sudan and South Sudan.

Involving international organisations can support national initiatives in Sudan to help contain the multiple factors driving the spiral of violence. Organisations like UNEP, the European Commission or the Nile Basin Initiative are encouraged to continue the dialogue with Sudanese authorities to identify joint initiatives for environmental and resource management to support peace building and development in Sudan and South Sudan.[87] They could provide a frame for an open and regular dialogue on Nile waters.

5. Conclusion: integration and policy implications

This article demonstrates the need for a broad and systematic understanding of the climate–conflict–violence nexus and its central concepts. Based on analysis of the mainly quantitative literature, we showed some blind spots in the current research on the link between climate change and violence. We then presented a theoretical framework and potential pathways between climate change and violent conflict that provide a basis for more systematic research. To understand the role of environmental factors in emerging spirals of violence, an agent-based approach is applied to assess the dynamic interaction between conflict parties and the conditions for conflicting or cooperative interactions.

Particular attention is paid to the concept of climate change as a possible ‘threat multiplier’, exceeding the adaptive capacities of communities and undermining their livelihoods. If actors react with violent strategies to secure or acquire resources, this could lead to insecurity and increased threat perceptions, driving a security dilemma that increasingly diverts resources into the spiral of violence. In the most affected states, the erosion of social order, state failure and pre-existent, ongoing violent conflicts could be aggravated, leading to a cycle of corruption, crime and behavioural violence that further dissolves state structures.

Against this background we investigate case studies in Kenya and Sudan, focusing on factors multiplying or preventing a spiral of violence. While we analysed an interpastoral conflict with a limited number of deaths in Kenya, the Sudan study represents a civil war between (state-controlled) militias, rebel forces and local people who have experienced tremendous numbers of casualties and refugees. While socio-economic marginalisation and access to arms are relevant in both cases, rent seeking of non-renewable resources (mainly oil) and geo-political forces are stronger motives in Sudan. Here, the impact of climate change is also less direct than in the case of Kenya. In both cases, it is essential to reduce political and economic marginalisation, develop resource-sharing mechanisms and maintain mobility for pastoralists. For Sudan this is particularly challenging, as Darfur is a strongly marginalised region and the devastating war allows for little development or adaptation measures.

To avoid climate-related violence in the future, it is important to develop an integrated framework and long-term strategies for a sustainable and peaceful mitigation of and adaptation to climate change. The following strategies can be suggested to support a transition to a more sustainable society and prevent instances of behavioural and structural violence related to climate change:

- Ambitious climate change mitigation strategies are required to limit the rise of the global mean temperature and its knock-on effects on resource availability and structural violence. Particularly important are mitigation strategies transferring the basis of capitalism and globalisation from a carbon-intensive energy system to one composed of renewable energies such as wind, solar and hydropower.
- New capabilities are needed to manage disasters, including emergency planning and decision-making structures. Global information systems for early warning could help with timely responses to extreme events.
- Adaptation strategies allow societies to develop technological, physical, human and social capital to sustain human livelihoods under changing climatic conditions, such as utilising natural resources more efficiently, growing and producing new types of natural resources and providing sustainable energy supply.
- Human migration is an adaptive response to climate change that is not only a threat, but could also offer opportunities.[88] Migrant networks can facilitate the exchange of knowledge, income and other resources across regions to strengthen adaptive capacities and resilience against climate change.[89]
- Regional security concepts would reduce the motivation to use violence by crisis prevention, conflict resolution and common security to stabilise fragile and weak states threatened by climate change. This is accompanied by mechanisms to restrain violent forces in conflicts by arms control, non-proliferation and disarmament.
- Innovative social mechanisms and institutional processes, including participative concepts, stakeholder dialogues, mediation, and adaptive governance, support people in regions affected by climate change in their creative efforts for protection of livelihoods, joint risk management and fair sharing of resources.
- The feasibility of recommendations and strategies dealing with climate change is to be based on the analysis of the drivers behind the economic determinants and geopolitical forces that shape globalisation and government responses to climate change, including the rise of neoliberalism and the ignorance of traditional institutions at the local level.

Acknowledgements

This research was partially founded by the German Environmental Foundation (DBU) and the Cluster of Excellence 'Integrated Climate System Analysis and Prediction – CliSAP' (EXC177), University of Hamburg, funded by the German Science Foundation (DFG). We thank Jürgen Zimmerer and two anonymous reviewers for their valuable comments.

Notes

1. For a definition see the glossary of the IPCC. http://www.ipcc.ch/publications_and_data/ar4/wg1/en/annex1sglossary-a-d.html): 'Climate in a narrow sense is usually defined as the average weather, or more rigorously, as the statistical description in terms of the mean and variability of relevant quantities over a period of time ranging from months to thousands or millions of years. The classical period for averaging these variables is 30 years, as defined by the World Meteorological Organization. The relevant quantities are most often surface variables such as temperature, precipitation and wind.' 'Climate change refers to a change in the state of the climate that can be identified (for example, by using statistical tests) by changes in the mean and/or the variability of its properties, and that persists for an extended period, typically decades or longer. Climate change may be due to natural internal processes or external forcings, or to persistent anthropogenic changes in the composition of the atmosphere or in land use.'
2. WBGU (German Advisory Council on Global Change), *World in Transition – Climate Change as a Security Risk* (London: Earthscan, 2008).
3. Jürgen Scheffran, Michael Brzoska, Hans Günter Brauch, Peter Michael Link and Janpeter Schilling, ed., *Climate Change, Human Security and Violent Conflict: Challenges for Societal Stability* (Berlin: Springer, 2012); Hans Günter Brauch, 'Securitizing Global Environmental Change', in *Facing Global Environmental Change: Environmental, Human, Energy, Food, Health and Water Security Concepts*, ed. Hans Günter Brauch, Úrsula Oswald Spring, John Grin, Czeslaw Mesjasz, Patricia Kameri-Mbote, Navnita Chadha Behera, Béchir Chourou and Heinz Krummenacher (Berlin; Heidelberg; New York: Springer, 2009), 65–102; Jürgen Scheffran and Antonella Battaglini, 'Climate and Conflicts: The Security Risks of Global Warming', *Regional Environmental Change* 11 (2011): 27–39.
4. Jürgen Scheffran, Michael Brzoska, Jasmin Kominek, P. Michael Link and Janpeter Schilling, 'Climate Change and Violent Conflict', *Science* 336 (2012): 869–71; Nils Petter Gleditsch, 'Special Issue on Climate Change and Conflict', *Journal of Peace Research* 49 (2012): 3–257.
5. Thorsten Bonacker,'Konflikttheorien', in *Handbuch Soziologische Theorien*, ed. Georg Kneer and Markus Schroer (Wiesbaden: VS-Verlag, 2009), 179–97.
6. Douglas P. Fry and Kay Björkqvist, *Cultural Variation in Conflict Resolution: Alternatives to Violence* (Mahwah, NJ: Lawrence Erlbaum Publishers, 1997).
7. Etienne Krug, Linda Dahlberg, James Mercy, Anthony Zwi and Rafael Lozano, *World Report on Violence and Health* (Geneva: World Health Organization, 2002).
8. Nils Petter Gleditsch, Peter Wallensteen, Mikael Eriksson, Margareta Sollenberg and Håvard Strand, 'Armed Conflict 1946–2001: A New Dataset', *Journal of Peace Research* 39 (2002): 615–37.
9. U.N. General Assembly, 'Convention on the Prevention and Punishment of the Crime of Genocide'. http://www.historyplace.com/worldhistory/genocide/genocide-convention.htm.

10. John Galtung, 'Violence, Peace, and Peace Research', *Journal of Peace Research* 6 (1969): 167–91.
11. Patrick Wolfe, 'Settler Colonialism and the Elimination of the Native', *Journal of Genocide Research* 8 (2006): 387–409.
12. Jennifer Huseman and Damien Short, 'A Slow Industrial Genocide: Tar Sands and the Indigenous Peoples of Northern Alberta', *The International Journal of Human Rights* 16 (2012): 216–37.
13. Ole Magnus Theisen, Nils Petter Gleditsch and Halvard Buhaug, 'Is Climate Change a Driver of Armed Conflict?', *Climatic Change* 117 (2013): 613–25; Jürgen Scheffran and Antonella Battaglini, 'Climate and Conflicts: The Security Risks of Global Warming', *Regional Environmental Change* 11 (2011): 27–39. See also section 'Climate-driven war and genocide?' in Gregory Kent, 'Crystallisations of the Global Western State in the Era of Climate Change', *The International Journal of Human Rights* 18, no. 3 (2014): 320–35.
14. Jon Barnett, 'Security and Climate Change', *Global Environmental Change-Human and Policy Dimensions* 13 (2003): 7–17; Thomas Homer-Dixon, 'Environmental Scarcities and Violent Conflict – Evidence from Cases', *International Security* 19 (1994): 5–40.
15. Betsy Hartmann, 'Rethinking Climate Refugees and Climate Conflict: Rhetoric, Reality and the Politics of Policy Discourse', *Journal of International Development* 22 (2010): 233–46; Cord Jakobeit and Chris Methmann, '"Climate Refugees" as Dawning Catastrophe? A Critique of the Dominant Quest for Numbers', in *Climate Change, Human Security and Violent Conflict: Challenges for Societal Stability*, ed. Jürgen Scheffran, Michael Brzoska, Hans Günter Brauch, Peter Michael Link and Janpeter Schilling (Berlin: Springer, 2012), 301–14.
16. Günther Bächler, 'Why Environmental Transformations Causes Violence: A Synthesis', *Environmental Change and Security Project Report* 4 (1998): 24–44; Oli Brown, Anne Hammill and Robert McLeman, 'Climate Change as the "New" Security Threat: Implications for Africa', *International Affairs* 83 (2007): 1141–54; Idean Salehyan, 'From Climate Change to Conflict? No Consensus Yet', *Journal of Peace Research* 45 (2008): 315–26; WBGU, *World in Transition*.
17. Rafael Reuveny, 'Climate Change-Induced Migration and Violent Conflict', *Political Geography* 26 (2007): 656–73.
18. See Appendix I for an overview.
19. Gleditsch et al., 'Armed Conflict 1946–2001'.
20. Colin H. Kahl, *States, Scarcity, and Civil Strife in the Developing World* (Princeton, NJ: Princeton University Press, 2006).
21. Janpeter Schilling, Francis Opiyo and Jürgen Scheffran, 'Raiding Pastoral Livelihoods: Motives and Effects of Violent Conflict in North-western Kenya', *Pastoralism* 2 (2012): 1–16; Wario Adano, Ton Dietz, Karen M. Witsenburg and Fred Zaal, 'Climate Change, Violent Conflict and Local Institutions in Kenya's Dryland', *Journal of Peace Resarch* 49 (2012): 65–80.
22. Hanne Fjelde and Nina von Uexkull, 'Climate Triggers: Rainfall Anomalies, Vulnerability and Communal Conflict in Sub-Saharan Africa', *Political Geography* 31 (2012): 444–53; Ole Magnus Theisen, 'Climate Clashes? Weather Variability, Land Pressure, and Organized Violence in Kenya, 1989–2004', *Journal of Peace Research* 49 (2012): 81–96.
23. John O'Loughlin, Frank Witmer, Andrew Linke, Arlene Laing, Andrew Gettelman and Jimy Dudhia, 'Climate Variability and Conflict Risk in East Africa, 1990–2009', *PNAS* 109 (2012): 18344–9; WBGU, *World in Transition*.
24. Jürgen Scheffran, P. Michael Link and Janpeter Schilling, 'Theories and Models of Climate-Security Interaction: Framework and Application to a Climate Hot Spot in North Africa', in *Climate Change, Human Security and Violent Conflict*, ed. Jürgen Scheffran, Hans Günter Brauch, Michael Brzoska, P. Michael Link and Janpeter Schilling (Berlin; Heidelberg: Springer, 2012), 91–131.
25. IPCC (Intergovernmental Panel on Climate Change), *Climate Change 2007: Climate Change Impacts, Adaptation and Vulnerability* (Geneva: Cambridge University Press, 2007); Hans-Martin Füssel, 'How Inequitable is the Global Distribution of Responsibility, Capability, and Vulnerability to Climate Change: A Comprehensive Indicator-Based Assessment', *Global Environmental Change* 20 (2011): 597–611. For a discussion of vulnerability to the power of states see Mark Levene and Daniele Conversi, 'Subsistence Societies, Globalisation, Climate Change and Genocide: Discourses of Vulnerability and Resilience', *The International Journal of Human Rights* 18, no. 3 (2014): 281–97.

26. Jon Barnett and W. Neil Adger, 'Climate Change, Human Security and Violent Conflict', *Political Geography* 26 (2007): 639–55.
27. The dimension of agricultural production and climate change in sub-Sahara Africa is discussed in more detail in Andreas Exenberger and Andreas Pondorfer, 'Genocidal Risk and Climate Change: Africa in the Twenty-First Century', *The International Journal of Human Rights* 18, no. 3 (2014): 350–68.
28. D.H. Smith, 'Poverty-Environment Linkages and their Implications for Security', in *Environmental Change and Human Security: Recognizing and Acting on Hazard Impacts*, ed. P.H. Liotta, David A. Mouat, William G. Kepner and Judith M. Lancaster (Dordrecht: Springer, 2008), 327–40.
29. Nancy Lee Peluso and Michael Watts, 'Violent Environments', in *Violent Environments*, ed. Nancy Lee Peluso and Michael Watts (Ithaca, NY; London: Cornell University Press, 2001), 3–38.
30. See for instance: CNA, *National Security and the Threat of Climate Change* (Alexandria: CNA, 2007); European Commission, *Climate Change and International Security: Paper from the High Representative and the European Commission to the European Council* (Brussels: European Commission, 2008).
31. Nicholas Sambanis, 'Using Case Studies to Expand Economic Models of Civil War', *Perspectives on Politics* 2 (2004): 259–79.
32. Sebastian Schutte and Nils B. Weidmann, 'Diffusion Patterns of Violence in Civil Wars', *Political Geography* 30 (2011): 143–52.
33. Jennifer Milliken and Keith Krause, 'State Failure, State Collapse and State Reconstruction: Concepts, Lessons, and Strategies', in *State Failure, Collapse and Reconstruction*, ed. Jennifer Milliken (London: Blackwell, 2003), 1–24; Ulrich Schneckener, 'States at Risk zur Analyse fragiler Staatlichkeit', in *States at Risk*, ed. Ulrich Schneckener (Berlin: Stiftung Wissenschaft und Frieden, 2004), 5–28; H. Starr, 'Failed States, Special Issue', *Conflict Management and Peace Science* 25 (2008).
34. Thomas Homer-Dixon, *Environmental Scarcity and Violence* (Princeton, NJ: Princeton University Press, 1999); Barnett and Adger, 'Climate Change, Human Security and Violent Conflict'.
35. Reuveny, 'Climate Change-Induced Migration and Violent Conflict'.
36. Philippe Le Billon, 'The Political Ecology of War: Natural Resources and Armed Conflicts', *Political Geography* 20 (2001): 561–84.
37. Michael Renner, *The Anatomy of Resource Wars* (Washington, DC: Worldwatch, 2002).
38. Jürgen Scheffran, Michael Brzoska, Jasmin Kominek, Michael Link and Janpeter Schilling, 'Disentangling the Climate-conflict Nexus: Empirical and Theoretical Assessment of Vulnerabilities and Pathways', *Review of European Studies* 4, no. 5 (2012): 1–13.
39. UNEP (United Nations Environment Programme), *From Conflict to Peacebuilding: The Role of Natural Resources and the Environment* (Nairobi: United Nations Environmental Programme, 2009).
40. http://www.thisisecoside.com (accessed 7 August 2013).
41. Charles Eisenstein, *The Ascent of Humanity* (Harrisburg: Panenthea Productions, 2007).
42. Jared Diamond, *How Societies Choose to Fail or Succeed* (New York: Penguin, 2011).
43. Polly Higgins, Damien Short and Nigel South, 'Protecting the Planet: A Proposal for a Law of Ecocide', *Crime, Law and Social Change* 59 (2013): 251–66.
44. Stephen Brosha, 'The Environment and Conflict in the Rwandan Genocide', *Atlantic International Studies Journal* 3 (2006). http://atlismta.org/online-journals/0607-journal-development-challenges/the-environment-and-conflict-in-the-rwandan-genocide/ (accessed 23 August 2013); Vadi Moodley, Alphonse Gahima and Suveshnee Munien, 'Environmental Causes and Impacts of the Genocide in Rwanda: Case Studies of the Towns of Butare and Cyangugu', *African Journal of Conflict Resolution* 10 (2010): 103–19.
45. Huseman and Short, 'A Slow Industrial Genocide'.
46. Higgins, Short and South, 'Protecting the Planet'.
47. Nicole Harari and Jesse Roseman, *Environmental Peacebuilding, Theory and Practice: A Case Study of the Good Water Neighbours Project and In Depth Analysis of the Wadi Fukin/Tzur Hadassah Communities* (Amman; Bethlehem; Tel Aviv: Friends of the Earth Middle East, 2008).
48. W. Neil Adger, 'Social Capital, Collective Action, and Adaptation to Climate Change', *Economic Geography* 79 (2003): 387–404.

49. Wolfgang Sofsky, *Zeiten des Schreckens: Amok, Terror, Krieg* (Frankfurt/Main: S. Fischer, 2002).
50. GoK (Government of Kenya), *National Policy for the Sustainable Development of Arid and Semi Arid Lands* (Nairobi: Government of Kenya, 2007).
51. Kennedy Mkutu, *Guns and Governance in the Rift Valley Pastoralist Conflict and Small Arms* (Bloomington: Indiana University Press, 2008).
52. David Eaton, *Violence, Revenge and the History of Cattle Raiding Along the Kenya-Uganda Border* (Halifax: Dalhousie University, 2008).
53. UNDP (United Nations Development Programme), *Drought and Potential Conflict Scenarios in Northern Kenya and Other Arid Lands: A Situational Report* (Nairobi: UNDP, 2011); Andrew Mude, Christopher B. Barrett, Michael R. Carter, Sommarat Chantarat, Munenobu Ikegami and John McPeak, *Index Based Livestock Insurance for Northern Kenya's Arid and Semi-Arid Lands: The Marsabit Pilot* (Nairobi: International Livestock Research Institute, 2010).
54. Eaton, *Violence, Revenge and the History of Cattle Raiding Along the Kenya-Uganda Border.*
55. Manasseh Wepundi, James Ndung'u and Simon Rynn, *Lessons From the Frontiers – Civilian Disarmament in Kenya and Uganda* (Nairobi: Saferworld, 2011).
56. Janpeter Schilling, Francis Opiyo and Jürgen Scheffran, 'Raiding Pastoral Livelihoods: Motives and Effects of Violent Conflict in North-Western Kenya', *Pastoralism* 2 (2012): 1–16.
57. Schilling, Opiyo and Scheffran, 'Raiding Pastoral Livelihoods'.
58. Kennedy Mkutu, 'Complexities of Livestock Raiding in Karamoja', *Nomadic Peoples* 14 (2010): 87–105; George Kaimba, Bernard Njehia and Abdi Guliye, 'Effects of Cattle Rustling and Household Characteristics on Migration Decisions and Herd Size Amongst Pastoralists in Baringo District, Kenya', *Pastoralism: Research, Policy and Practice* 1 (2011): 1–16.
59. Mkutu, *Guns and Governance in the Rift Valley Pastoralist Conflict and Small Arms.*
60. Eaton, *Violence, Revenge and the History of Cattle Raiding Along the Kenya-Uganda Border.*
61. TUPADO (Turkana Pastoralist Organisation), *Turkana Pastoralist Organisation Incident Register 2000–2010*, XXX.
62. Janpeter Schilling, *On Rains, Raids and Relations: A Multimethod Approach to Climate Change, Vulnerability, Adaptation and Violent Conflict in Northern Africa and Kenya* (Hamburg: PhD thesis, 2012).
63. David Eaton, 'The Rise of the "Traider": The Commercialization of Raiding in Karamoja', *Nomadic Peoples* 14 (2010): 106–22; Saverio Krätli and Jeremy Swift, *Understanding and Managing Pastoral Conflict in Kenya* (Sussex: University of Sussex, 2003); Kennedy Agade Mkutu, 'Complexities of Livestock Raiding in Karamoja', *Nomadic Peoples* 14 (2010): 87–105.
64. Janpeter Schilling, Francis Opiyo and Jürgen Scheffran, 'Raiding Pastoral Livelihoods: Motives and Effects of Violent Conflict in North-Western Kenya', *Pastoralism* 2 (2012): 1–16.
65. http://www.tullowoil.com/index.asp?pageid=432 (2 November 2012).
66. Sara Pantuliano, 'Oil, Land and Conflict: The Decline of Misseriyya Pastoralism in Sudan', *Review of African Political Economy* 37 (2010): 7–23.
67. Mude et al., *Index Based Livestock Insurance for Northern Kenya's Arid and Semi-Arid Lands.*
68. Carol McSweeney, Mark New and Gil Lizcano, 'UNDP Climate Change Country Profiles – Kenya'. http://country-profiles.geog.ox.ac.uk/.
69. Ibid.
70. Schilling, *On Rains, Raids and Relations.*
71. Ibid.
72. United Nations Office for the Coordination of Humanitarian Affairs UNOCHA, 'Security in Mobility'. http://ochaonline.un.org/kenya/Advocacy/SecurityinMobility/tabid/6735/language/en-US/Default.aspx.
73. Schilling, Opiyo and Scheffran, 'Raiding Pastoral Livelihoods'.
74. Charles Chavunduka and Daniel W. Bromley, 'Climate, Carbon, Civil War and Flexible Boundaries: Sudan's Contested Landscape', *Land Use Policy* 28 (2011): 907–16.
75. UNEP (United Nations Environment Programme), *Sudan Post-Conflict Environmental Assessment* (Nairobi: United Nations Environment Programme, 2007); Ian A. Brown, 'Assessing Eco-Scarcity as a Cause of the Outbreak of Conflict in Darfur: A Remote Sensing Approach', *International Journal of Remote Sensing* 31, no. 10 (2010): 2513–20.
76. Linda Dahlberg, *Sudan Environmental Policy Brief* (Gothenburg: Gothenburg University, Environmental Economic Unit, 2007).

77. International Commission of Inquiry on Darfur, *Report of the International Commission of Inquiry on Darfur to the United Nations Secretary-General* (Geneva: UN, 2005).
78. Alex de Waal and Y. Ajawin, ed., *Facing Genocide: The Nuba of Sudan* (London: African Rights, 1995); Abdelmoneim Hashim Elnagheeb and Daniel W. Bromley, 'Rainfed Mechanized Farming and Deforestation in Central Sudan', *Environmental and Resource Economics* 2 (1992): 359–71; M. Suliman, 'Civil War in the Sudan: From Ethnic to Ecological Conflict', *The Ecologist* 23 (1993): 104–9.
79. B. Swallow and D. Bromley, 'Institutions, Governance and Incentives in Common Property Regimes for African Rangelands', *Environmental and Resource Economics* 6 (1995): 99–118.
80. Julie Flint and Alex de Waal, *Darfur: A New History of a Long War* (London; New York: Zed Books, 2008).
81. Alex de Waal, *Sudan: International Dimensions to the State and its Crisis* (London: Crisis States Research Centre, 2007); Khalid Y. Khalafalla, 'Der Konflikt in Darfur', *Aus Politik und Zeitgeschichte* (2005): 40–6.
82. UNEP, *Sudan Post-Conflict Environmental Assessment*; Brown, 'Assessing Eco-Scarcity'.
83. Declan Butler, 'Darfur's Climate Roots Challenged', *Nature* 447 (2007): 1038.
84. Harry Verhoeven, 'Climate Change, Conflict and Development in Sudan: Global Neo-Malthusian Narratives and Local Power Struggles', *Development and Change* 42 (2011): 679–707; Marcel Leroy, *Environment and Conflict in Africa – Reflections on Darfur* (Addis Ababa, Ethiopia: University for Peace, 2009).
85. Dahlberg, *Sudan Environmental Policy Brief*.
86. Ibid.
87. Ibid.
88. R. Black, D. Kniveton and K. Schmidt-Verkerk, 'Migration and Climate Change: Towards an Integrated Assessment of Sensitivity', *Environment and Planning A* 43 (2011): 431–50.
89. Jürgen Scheffran, Elina Marmer and Papa Sow, 'Migration as a Contribution to Resilience and Innovation in Climate Adaptation: Social Networks and Co-Development in Northwest Africa', *Applied Geography* 33 (2012): 119–27.

Appendix 1

No.	Author(s)	Definition
1	Bergholt/Lujala (2012)[a]	UCDP/PRIO armed conflict
2	Biermann et al. (1998)[b]	Violent conflict incidence (KOSIMO)
3	Brückner/Ciccone (2010)[c]	UCDP/PRIO armed conflict
4	Buhaug (2010)[d]	UCDP/PRIO armed conflict Civil war
5	Buhaug/Theisen (2012)[e]	UCDP/PRIO armed conflict
6	Burke et al. (2009)[f]	UCDP/PRIO civil war
7	Drury/Olson (1998)[g]	Political unrest incidence (WHSI)
8	Esty et al. (1999)[h]	State failure events
9	Fjelde/van Uexkull (2012)[i]	UCDP/PRIO non-state conflict
10	Gizelis/Wooden (2010)[j]	UCDP/PRIO armed conflict
11	Hauge/Ellingsen (1998)[k]	UCDP/PRIO armed conflict
12	Hendrix/Glaser (2007)[l]	UCDP/PRIO armed conflict
13	Hendrix/Salehyan (2012)[m]	UCDP/PRIO armed conflict Violent conflict (SCAD) Individual battles (ACLED)
14	Hsiang/Meng/Cane (2011)[n]	UCDP/PRIO armed conflict
15	Jensen/Gleditsch (2009)[o]	UCDP/PRIO armed conflict
16	Koubi et al. (2012)[p]	UCDP/PRIO armed conflict
17	Levy et al. (2005)[q]*	UCDP/PRIO armed conflict
18	Melander/Sundberg (2011)[r] *	UCDP/Non-state conflict
19	Miguel/Satyanath/Sergenti (2004)[s]	UCDP/PRIO armed conflict
20	Nel/Righarts (2008)[t]	UCDP/PRIO armed conflict
21	O'Loughlin et al. (2012)[u]	Armed conflict events (ACLED)
22	Omelicheva (2011)[v]	Episodes of political instability (PITF)
23	Raleigh/Kniveton (2012)[w]	Armed conflict events (ACLED)
24	Raleigh/Urdal (2007)[x]	UCDP/PRIO armed conflict
25	Rowhani et al.(2011)[y]	UCDP/PRIO armed conflict
26	Slettebak (2012)[z]	UCDP/PRIO armed conflict
27	Theisen (2008)[aa]	UCDP/PRIO armed conflict
28	Theisen (2012)[bb]	Intergroup violence
29	Theisen/Brandsegg (2007)[cc]*	UCDP/Non-state conflict
30	Theisen/Holtermann/Buhaug (2012)[dd]	UCDP/PRIO armed conflict
31	Tol/Wagner (2010)[ee]	Wars that last longer than one year and are significant enough to be recognised by historians (original source www.warscholar.com does not exist anymore)
32	Urdal (2005)[ff]	UCDP/PRIO armed conflict
33	Urdal (2008)[gg]	UCDP/PRIO armed conflict Political violence events (Keesing's Record of World Events) Hindu–Muslim riots
34	Wischnath/Buhaug (2012)[hh]*	PRIO Conflict Site Dataset
35	Zhang et al. (2006)[ii]	Wars
36	Zhang et al. (2007)[jj]	Violent conflict Wars
37	Zhang et al. (2007)[kk]	Wars
38	Zhang et al. (2010)[ll]	Wars

(*Continued*)

Appendix 1. Continued

No.	Author(s)	Definition
39	Zhang et al. (2011)[mm]	Violent conflict

Notes: *Has not yet been published in a peer-reviewed journal.

[a]Drago Bergholt and Päivi Lujala, 'Climate-Related Natural Disasters, Economic Growth, and Armed Civil Conflict', *Journal of Peace Research* 49 (2012): 147–62.
[b]Frank Biermann, Gerhard Petschel-Held and Christian Rohloff, 'Umweltzerstörung als Konfliktursache? Theoretische Konzeptionalisierung und empirische Analyse des Zusammenhangs von "Umwelt" und "Sicherheit"', *Zeitschrift für Internationale Beziehungen* 5 (1998): 273–308.
[c]Markus Brückner and Antonio Ciccone, 'International Commodity Prices, Growth, and the Outbreak of Civil War in Sub-Saharan Africa', *The Economic Journal* 120 (2010): 519–34.
[d]Halvard Buhaug, 'Climate Not to Blame for African Civil Wars', *PNAS* 107 (2010): 16477–82.
[e]Halvard Buhaug and Ole Magnus Theisen, 'On Environmental Change and Armed Conflict', in *Climate Change, Human Security and Violent Conflict: Challenges for Societal Stability*, ed. Jürgen Scheffran, Michael Brzoska, Hans Günter Brauch, Peter Michael Link and Janpeter Schilling (Berlin: Springer, 2012), 43–55.
[f] Marshall Burke, Edward Miguel, Shanker Satyanath, John Dykema and David Lobell, 'Warming Increases the Risk of Civil War in Africa', *Proceedings of the National Academy of Sciences* 106 (2009): 20670–4.
[g]A. Cooper Drury and Richard Stuart Olson, 'Disasters and Political Unrest: An Empirical Investigation', *Journal of Contingencies and Crisis Management* 6 (1998): 153–61.
[h]Daniel Esty, Jack Goldstone, Ted Robert Gurr, Barbara Harff, Marc Levy, Geoffrey Dabelko, Pamela Surko and Allan Unge, 'State Failure Task Force Report: Phase II Findings', *Environmental Change and Security Project Report* 5 (1999): 49–72.
[i]Hanne Fjelde and Nina von Uexkull, 'Climate Triggers: Rainfall Anomalies, Vulnerability and Communal Conflict in Sub-Saharan Africa', *Political Geography* 31 (2012): 444–53.
[j]Theodora-Ismene Gizelis and Amanda E. Wooden, 'Water Resources, Institutions, & Intrastate Conflict', *Political Geography* 29 (2010): 444–53.
[k]Wenche Hauge and Tanja Ellingsen, 'Beyond Environmental Scarcity: Causal Pathways to Conflict', *Journal of Peace Research* 35 (1998): 299–317.
[l]Cullen S. Hendrix and Sarah M. Glaser, 'Trends and Triggers: Climate, Climate Change and Civil Conflict in Sub-Saharan Africa', *Political Geography* 26 (2007): 695–715.
[m]Cullen S. Hendrix and Idean Salehyan, 'Climate Change, Rainfall, and Social Conflict in Africa', *Journal of Peace Research* 49 (2012): 35–50.
[n]Solomon M. Hsiang, Kyle C. Meng and Mark A. Cane, 'Civil Conflicts are Associated with the Global Climate', *Nature* 476 (2011): 438–41.
[o]Peter Sandholt Jensen and Kristian Skrede Gleditsch, 'Rain, Growth, and Civil War: The Importance of Location', *Defense and Peace Economics* 20 (2009): 359–72.
[p]Vally Koubi, Thomas Bernauer, Anna Kalbhenn and Gabriele Spilker, 'Climate Variability, Economic Growth, and Civil Conflict', *Journal of Peace Research* 49 (2012): 113–27.
[q]M.A. Levy, C. Thorkelson, C. Vörösmarty, E. Douglas and M. Humphreys, 'Freshwater Availability Anomalies and Outbreak of Internal War: Results from a Global Spatial Time Series Analysis' (paper presented at Human Security and Climate Change, Oslo, Norway, 21–23 June 2005).
[r]Erik Melander and Ralph Sundberg, 'Climate Change, Environmental Stress, and Violent Conflict – Tests Introducing the UCDP Georeferenced Event Dataset' (paper presented at 51st Annual Convention of the International Studies Association Montreal, 2011).
[s]Edward Miguel, Shanker Satyanath and Ernest Sergenti, 'Economic Shocks and Civil Conflict: An Instrumental Variables Approach', *Journal of Political Economy* 112 (2004): 725–53.
[t]Philip Nel and Marjolein Righarts, 'Natural Disasters and the Risk of Violent Civil Conflict', *International Studies Quarterly* 52 (2008): 159–85.
[u]John O'Loughlin, Frank Witmer, Andrew Linke, Arlene Laing, Andrew Gettelman and Jimy Dudhia, 'Climate Variability and Conflict Risk in East Africa, 1990–2009', *Proceedings of the National Academy of Sciences (PNAS)* 109 (2012): 18344–9.
[v]Mariya Y. Omelicheva, 'Natural Disasters: Triggers of Political Instability?', *International Interactions* 37 (2011): 441–65.
[w]Clionadh Raleigh and Dominic Kniveton, 'Come Rain or Shine: An Analysis of Conflict and Climate Variability in East Africa', *Journal of Peace Research* 49 (2012): 51–64.
[x]Clionadh Raleigh and Henrik Urdal, 'Climate Change, Environmental Degradation and Armed Conflict', *Political Geography* 26 (2007): 674–94.
[y]Pedram Rowhani, Olivier Degomme, Debarati Guha-Sapir and Eric Lambin, 'Malnutrition and Conflict in East Africa: The Impacts of Resource Variability on Human Security', *Climatic Change* 105 (2011): 207–22.

(*Continued*)

Appendix 1. Continued

[z]Rune T. Slettebak, 'Don't Blame the Weather! Climate-Related Natural Disasters and Civil Conflict', *Journal of Peace Research* 49 (2012): 163–76.
[aa]Ole Magnus Theisen, 'Blood and Soil? Resource Scarcity and Internal Armed Conflict Revisited', *Journal of Peace Research* 45 (2008): 801–18.
[bb]Ole Magnus Theisen, 'Climate Clashes? Weather Variability, Land Pressure, and Organized Violence in Kenya, 1989–2004', *Journal of Peace Research* 49 (2012): 81–96.
[cc]Ole Magnus Theisen and Kristian Bjarnøe Brandsegg, 'The Environment and Non-State Conflicts in Sub-Saharan Africa' (paper presented at 48th Annual Convention of the International Studies Association, Chicago, 28 February–3 March 2007).
[dd]Ole Magnus Theisen, Helge Holtermann and Halvard Buhaug, 'Climate Wars? Assessing the Claim That Drought Breeds Conflict', *International Security* 36 (2012): 79–106.
[ee]Richard Tol and Sebastian Wagner, 'Climate Change and Violent Conflict in Europe Over the Last Millennium', *Climatic Change* 99 (2010): 65–79.
[ff]Henrik Urdal, 'People vs. Malthus: Population Pressure, Environmental Degradation, and Armed Conflict Revisited', *Journal of Peace Research* 42 (2005): 417–34.
[gg]Henrik Urdal, 'Population, Resources, and Political Violence', *Journal of Conflict Resolution* 52 (2008): 590–617.
[hh]Gerdis Wischnath and Halvard Buhaug, 'Heat Waves and Rebel Raves? On Climate Variability and Civil War in Asia' (paper presented at 52nd Annual Convention of the International Studies Association, San Diego, 2012).
[ii]D.D. Zhang, C.Y. Jim, G.C.S. Lin, Y.Q. He, J.J. Wang and H.F. Lee, 'Climatic Change, Wars and Dynastic Cycles in China Over the Last Millennium', *Climatic Change* 76 (2006): 459–77.
[jj]David D. Zhang, Peter Brecke, Harry F. Lee, Yuan-qing He and Jane Zhang, 'Global Climate Change, War and Population Decline in Recent Human History', *Proceedings of the National Academy of Sciences* 104 (2007): 19214–19.
[kk]David D. Zhang, Jane Zhang, Harry F. Lee and Yuan-qing He, 'Climate Change and War Frequency in Eastern China over the Last Millennium', *Human Ecology* 35 (2007): 403–14.
[ll]Z.-H. Zhang, Helen Tian, Bernard Cazelles, Kyrre L. Kausrud, Achim Brauning, Fh Guo and Nils Chr. Stenseth, 'Periodic Climate Cooling Enhanced Natural Disasters and Wars in China During AD 10–1910', *Proceedings of the Royal Society B: Biological Sciences* 277 (2010): 3745-3753.
[mm]David D. Zhang, Harry F. Lee, Cong Wang, Baosheng Li, Qing Pei, Jane Zhang and Yulun An, 'The Causality Analysis of Climate Change and Large-Scale Human Crisis', *Proceedings of the National Academy of Sciences* 108 (2011): 17296–301.

Index

Page numbers followed by "n" indicate notes. Page numbers in bold indicate tables, and in italics indicate figures.

For Product Safety Concerns and Information please contact our EU representative GPSR@taylorandfrancis.com Taylor & Francis Verlag GmbH, Kaufingerstraße 24, 80331 München, Germany

Batch number: 10726294

Printed by Printforce, the Netherlands